SIMPLY
GENETICS

CONSULTANT AND CONTRIBUTOR

Derek Harvey is a naturalist with a particular interest in evolutionary biology. He graduated in zoology from Liverpool University in the UK and taught a generation of biologists, leading student expeditions to Costa Rica, Madagascar, and Australasia. Derek now concentrates on writing and consulting for science and natural history books.

CONTRIBUTOR

Jo Locke is an accomplished educational consultant and author. She studied biology at the University of Bath and is known for her contributions to science education, particularly in developing engaging and effective learning resources for students.

CONTENTS

7 **WHAT IS GENETICS?**

THE GENETIC LANDSCAPE

10 **GENES AND LIFE**
Living vs non-living things

11 **A MATTER OF BREEDING**
Genes and species

12 **THE DIVERSITY OF LIFE**
Genes and variety

14 **UNITS OF LIFE**
Cells

16 **THE INFORMATION CENTRE**
Genes and chromosomes

17 **FROM GENES TO PROTEINS**
Controlling cell activity

18 **MORE OR LESS**
Chromosome numbers

19 **A MATTER OF SEX**
Sex chromosomes

20 **CHROMOSOME PATTERNS**
Karyotypes

22 **THE CONTINUITY OF LIFE**
Types of reproduction

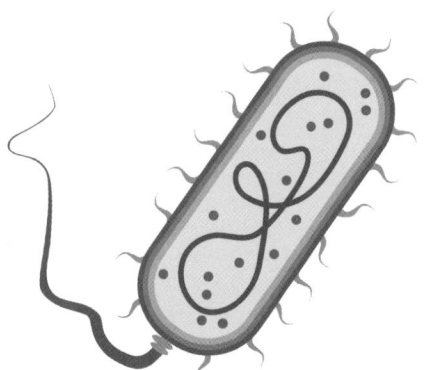

24 **HALF AND HALF**
Chromosomes and sex cells

25 **VARIANTS OF GENES**
Alleles

26 **ALLELE INTERACTIONS**
Dominant or recessive

40 **FINDING THE FACTORS**
The chromosome theory of inheritance

42 **TIED TOGETHER**
Linked genes

43 **GENE GEOGRAPHY**
Linkage maps

44 **MENDEL'S FORTUNE**
The physical basis of inheritance

VARIATION

48 **GENOTYPE AND PHENOTYPE**
Describing variation

49 **STEPPED STATES**
Discrete variation

50 **GENES IN COMBINATION**
Alleles and populations

51 **A SLIDING SCALE**
Continuous variation

52 **GENE INTERACTION**
Epistasis

53 **OUTSIDE INFLUENCES**
Environmental variation

INVESTIGATING INHERITANCE

30 **NESTED LIFE?**
Preformation

31 **GENETIC FACTORS?**
Material inheritance

32 **LIFE IN THE MIX**
Mechanics of inheritance

34 **GENETIC TEST BEDS**
Model organisms

36 **3:1 INHERITANCE**
Simple crosses

38 **RATIOS OF INHERITANCE**
Dihybrid crosses

THE DNA MOLECULE

56 **LADDERS OF LIFE**
   Molecules of inheritance
58 **A TWIST OF FATE**
   The double helix
60 **MEMORY MOLECULES**
   Templates and replication
62 **PACKED IN**
   DNA coiling and packaging
63 **CELLULAR COURIERS**
   Ribonucleic acid
64 **THE 98 PER CENT**
   Non-coding DNA
65 **NON-NUCLEAR DNA**
   Mitochondrial DNA
66 **LIFE IN THE ROUND**
   The cell cycle
67 **DOUBLING DOWN**
   Types of cell division
68 **DANCE OF THE CHROMOSOMES**
   Phases of mitosis
70 **QUICK COPIES**
   DNA replication

GENES AND REPRODUCTION

74 **CHANGE FOR THE BETTER**
   Generating genetic diversity
76 **STATES OF BEING**
   Diploid and haploid
78 **REDUCTION DIVISION**
   Meiosis I
80 **FORMING GAMETES**
   Meiosis II

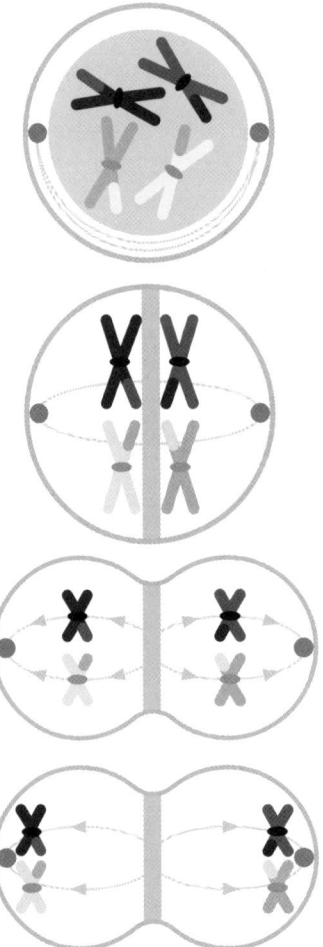

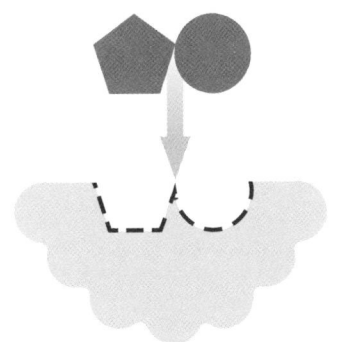

HOW GENES WORK

84 **ONE GENE, ONE PROTEIN**
Specific codes

85 **MOLECULES OF LIFE**
What proteins do

86 **FORM AND FUNCTION**
Protein structure

87 **BIOLOGICAL CATALYSTS**
Proteins and enzymes

88 **THE CENTRAL DOGMA**
The cell's information system

89 **THREE-LETTER WORDS**
Triplet code

90 **READING DNA**
Transcription

92 **BUILDING PROTEINS**
Translation

94 **MAKING THE CUT**
RNA editing

96 **ON AND OFF SWITCHES**
Gene regulation

98 **CONTROL SYSTEMS**
Genes in development

100 **GENE TAGGING**
Epigenetics

101 **CHEMICAL ASYMMETRY**
Drivers of early development

WHEN GENES GO WRONG

104 **COSTLY MISTAKES**
Causes of gene mutation

105 **WHEN MUTATIONS OCCUR**
Somatic and germline mutation

106 **COPYING ERRORS**
Gene mutations

107 **SHIFT ALONG**
Frameshift mutations

108 **SORTING ERRORS**
Whole chromosome mutation

109 **MOVING SECTIONS**
Structural chromosome mutation

110 **RESHAPING PROTEINS**
Mutations and gene function

112 **UNWANTED LEGACY**
Genetic diseases

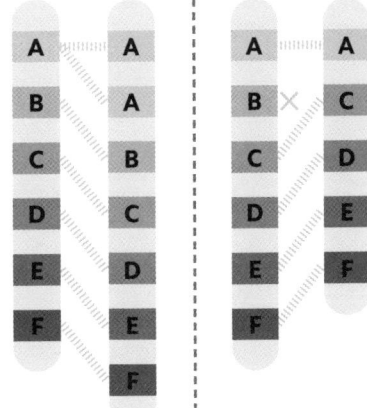

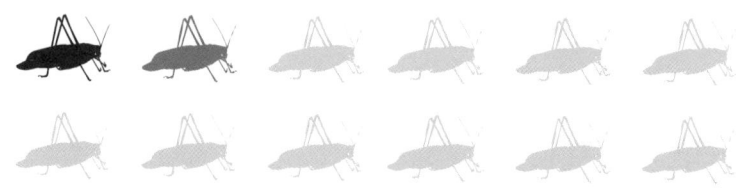

POPULATION GENETICS AND EVOLUTION

- 116 **GENES IN POPULATIONS**
  Individuals, populations, and species
- 117 **KEEPING STABLE**
  Genetic equilibrium
- 118 **AGENTS OF CHANGE**
  The mechanisms of evolution
- 120 **FORTUNATE MISTAKES**
  Evolution by mutation
- 121 **CHANCE CHANGE**
  Bottlenecks and founders
- 122 **BEST FIT**
  Natural selection
- 124 **POPULATION PRESSURES**
  Types of selection
- 126 **THE MATING GAME**
  Sexual selection
- 127 **SCALE IN EVOLUTION**
  Micro- and macroevolution
- 128 **SPLITTING UP**
  Allopatric speciation
- 130 **EVOLVING TOGETHER**
  Sympatric speciation
- 131 **KEEPING APART**
  Reproductive barriers
- 132 **DYING OUT**
  Genetics and extinction
- 133 **SELFISH GENES**
  Units of selection

MANIPULATING GENES

- 136 **PROTEIN SEQUENCING**
  Analysing genes via proteins
- 137 **FINDING BAD GENES**
  Medical gene tests
- 138 **MAKING A MATCH**
  DNA fingerprinting
- 140 **CLONING DNA**
  Polymerase chain reaction (PCR)
- 142 **BASE TO BASE**
  DNA sequencing
- 144 **MAPPING DIVERSITY**
  The Human Genome Project
- 145 **PUTTING GENETICS TO WORK**
  Engineering living things
- 146 **MICROBE MECHANICS**
  Engineering bacteria
- 148 **CARRIERS OF CHANGE**
  Gene vectors
- 150 **PASSING IT ON**
  Somatic and germline modification
- 151 **HEALING GENES**
  Gene therapy
- 152 **REWRITING GENES**
  CRISPR
- 154 **DE-EXTINCTION**
  Cloning and resurrection

- 156 **INDEX**

WHAT IS GENETICS?

The living world is as remarkable for its unity as it is for its diversity. Organisms as different as trees and truffles, microbes and monkeys share a common heritage that is revealed down a microscope and through chemical analysis. They all have cells that contain DNA – the famous double-helix molecule that makes us what we are. The key is in its fine structure, which is a form of coded information capable of specifying the characteristics of each living organism.

In 1909, the Danish biologist Wilhelm Johannsen coined the term '"gene" (originally derived from the Greek *genos*, meaning "origin") to describe any chemical factor that determines an inherited trait. This was nearly half a century after an Austrian friar called Gregor Mendel suggested that such factors could explain patterns of inheritance in pea plants, and a half-century before scientists in Cambridge and London found the physical basis of genes in DNA. Johannsen's word became the root of a new branch of biology focused on inheritance: genetics.

Today, geneticists continue the work of those early pioneers in the fields of evolutionary biology, medicine, agriculture, bioengineering, and more. They study the aspects of genetics that are broadly laid out in this book – from the principles and patterns of heredity and the building blocks of biological variety, to the basis of change by development and evolution. And as they advance humanity's understanding of the living world, geneticists equipped with modern technology can go further than ever before – by manipulating genes to cure disease, engineering organisms and their products, and even getting close to resurrecting extinct life.

THE GENET LANDS

Genetics is the biological study of inherited characteristics. It investigates how traits get passed down through generations and how those traits account for the diversity of living things. From molecules and cells to bodies and populations, genetics explains how microscopic heritable "factors", or genes, determine the physical make-up and behaviour of organisms, as well as explaining the differences between individuals and species. Molecular mechanisms underpin the continuity of life: genes are made up of a remarkable substance called DNA (deoxyribonucleic acid) that governs the growth, reproduction, and evolution of organisms.

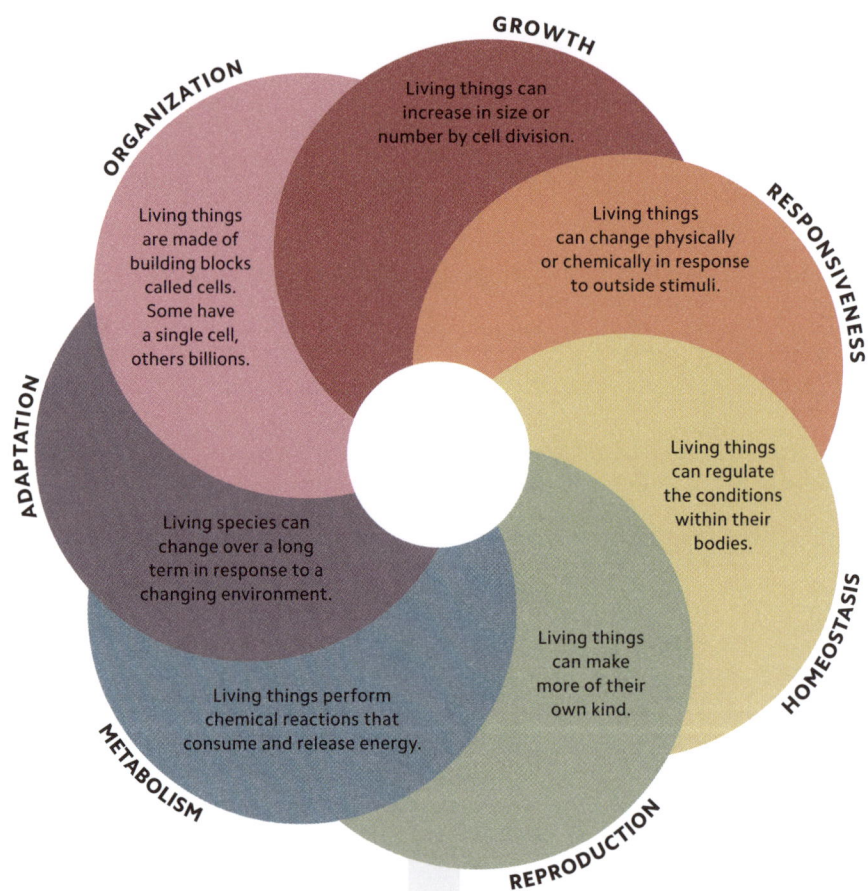

GENES AND LIFE

Livings things possess a suite of characteristics that are absent from non-living things. Above all, they are able to reproduce and pass their physical, biochemical, and behavioural traits to their offspring. This requires information – encoded by genes within every cell of an organism's body – to be passed from parent to offspring. This information may be altered through the generations, allowing life to adapt to changes in the environment.

A MATTER OF BREEDING

An individual organism carries a set of genes (or genome) that governs its characteristics. That organism is able to breed successfully only with others that share (or largely share) the same set of genes. Organisms that are able to interbreed and produce fertile offspring are said to be members of the same biological species. Closely related species can sometimes interbreed, but their offspring are often sterile.

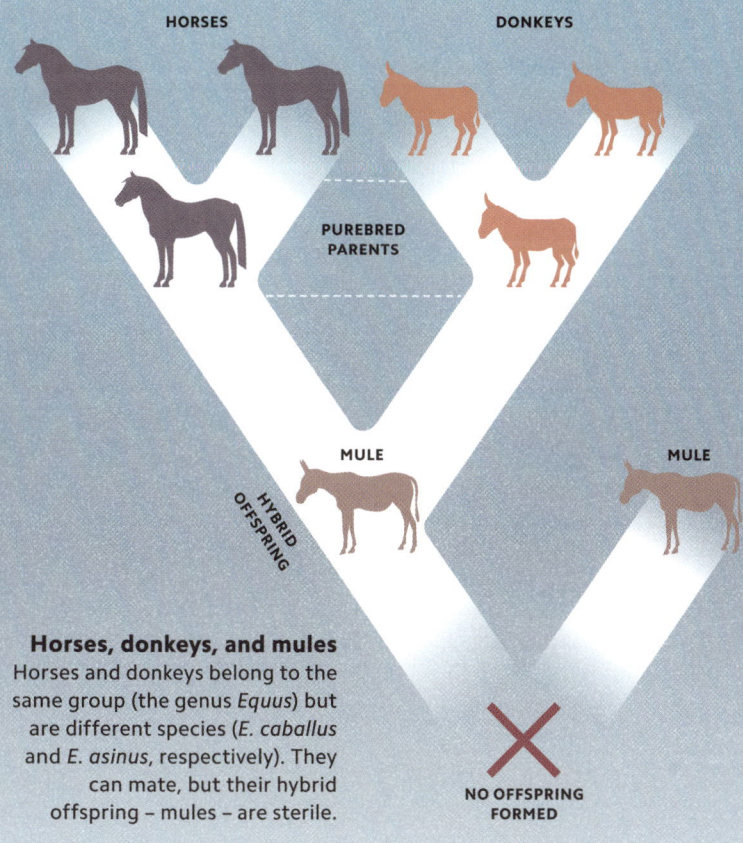

Horses, donkeys, and mules
Horses and donkeys belong to the same group (the genus *Equus*) but are different species (*E. caballus* and *E. asinus*, respectively). They can mate, but their hybrid offspring – mules – are sterile.

THE DIVERSITY OF LIFE

Scientists have named more than two million living species, and an estimated 10 million or more remain to be described. Species can be classified into one of six kingdoms that reflect key differences in their biology. The six kingdoms encompass a huge diversity of sizes, shapes, and ways of life – from bacteria less than a thousandth of a millimetre across to giant sperm whales and trees. The diversity of life stems from the properties of genetic material, which allows for evolutionary change.

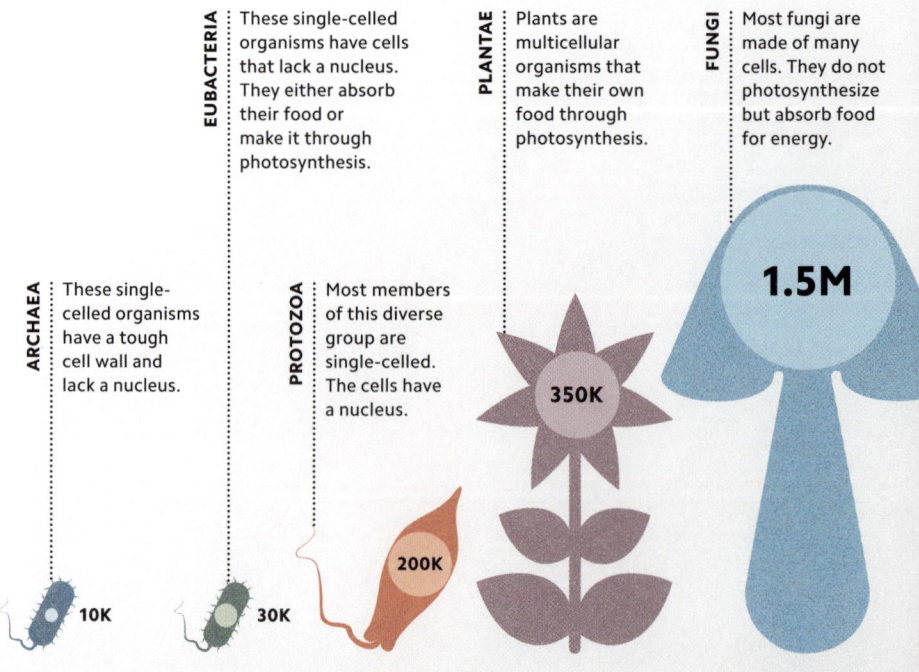

ARCHAEA — These single-celled organisms have a tough cell wall and lack a nucleus. 10K

EUBACTERIA — These single-celled organisms have cells that lack a nucleus. They either absorb their food or make it through photosynthesis. 30K

PROTOZOA — Most members of this diverse group are single-celled. The cells have a nucleus. 200K

PLANTAE — Plants are multicellular organisms that make their own food through photosynthesis. 350K

FUNGI — Most fungi are made of many cells. They do not photosynthesize but absorb food for energy. 1.5M

Kingdoms
The estimated numbers of living species in each kingdom reflect how deeply each one has been studied. Some biologists suggest that there may be in excess of two billion species.

9 MILLION

> Living things are a tiny fraction of biological diversity. Over 99 per cent of all plant and animal species that have lived are extinct.

ANIMALIA
Animals are multicellular organisms with nerves and muscles. They consume food.

GENES AND VARIETY | 13

UNITS OF LIFE

All living things are made from basic building blocks called cells. Complex organisms, such as humans, are composed of trillions of cells, while simple ones, such as bacteria and protozoa (see pp.12–13), may exist as a single cell. Cells provide the body with structure, take in and process food, and generate the energy to power processes such as movement, growth, development, and replication. Every cell contains DNA – the hereditary material that governs the functions of the cell and allows it to reproduce itself.

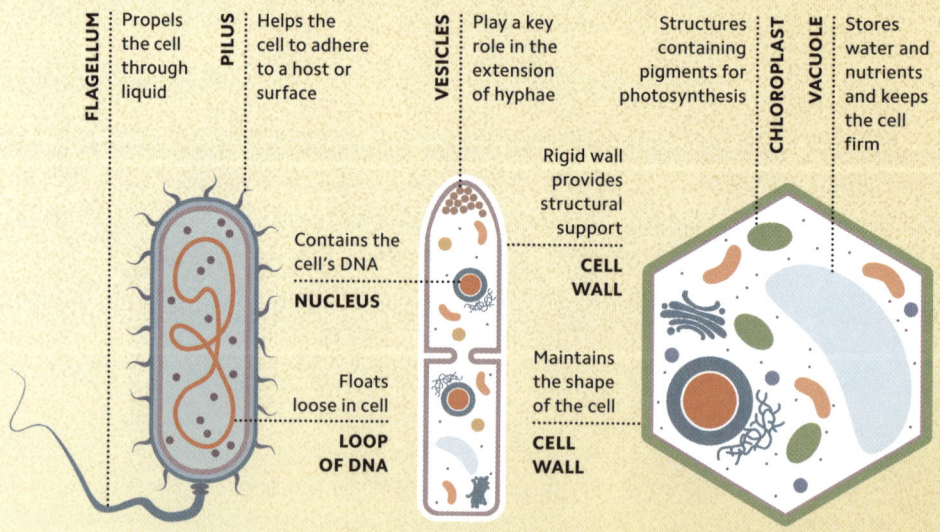

FLAGELLUM Propels the cell through liquid

PILUS Helps the cell to adhere to a host or surface

VESICLES Play a key role in the extension of hyphae

CHLOROPLAST Structures containing pigments for photosynthesis

VACUOLE Stores water and nutrients and keeps the cell firm

Contains the cell's DNA
NUCLEUS

Rigid wall provides structural support
CELL WALL

Floats loose in cell
LOOP OF DNA

Maintains the shape of the cell
CELL WALL

Bacterial cell
Bacteria are prokaryotes; their cells lack a nucleus and are simpler in structure than other cells.

Fungal cell
The cells of fungi form long filaments called hyphae. They release enzymes that digest organic materials.

Plant cell
These cells have rigid cell walls made of cellulose. They are able to produce their own food by photosynthesis.

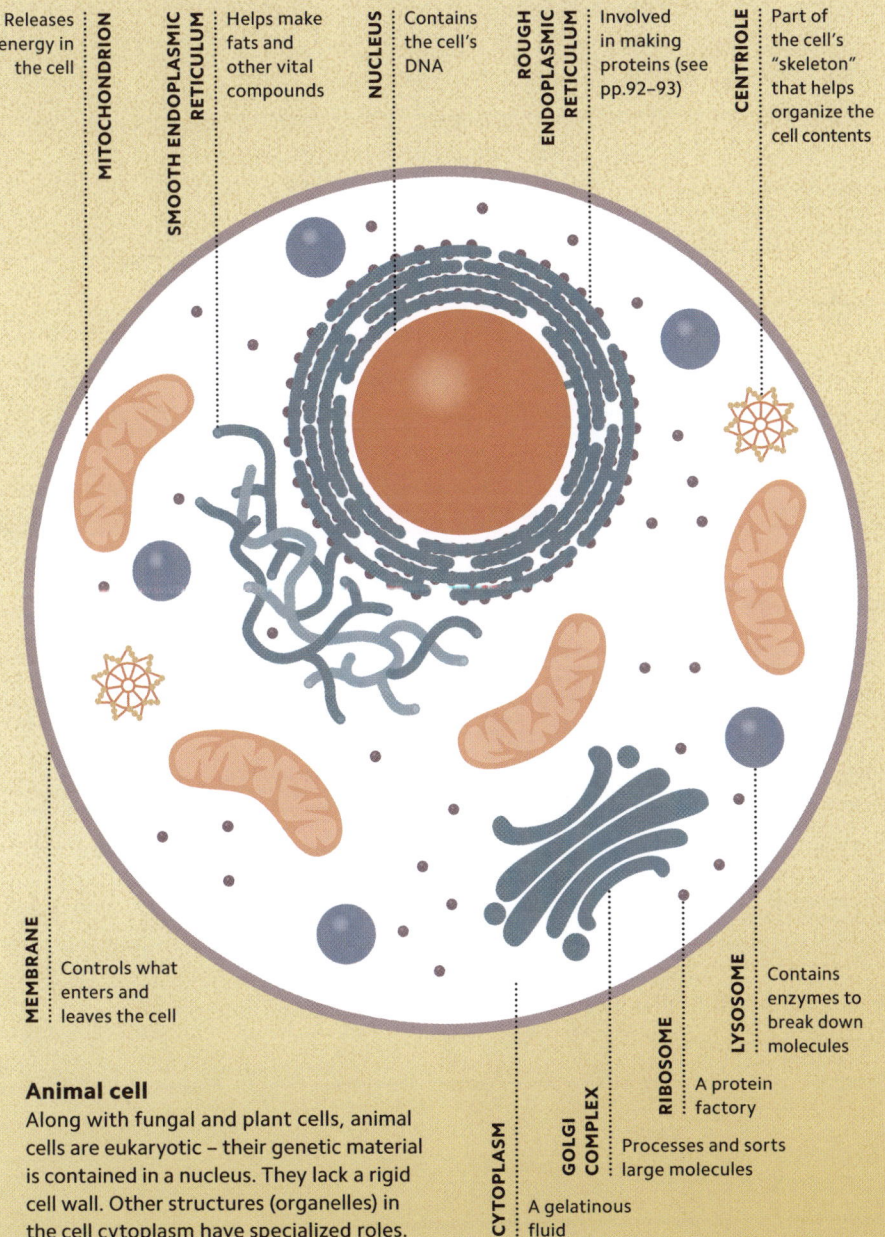

Animal cell
Along with fungal and plant cells, animal cells are eukaryotic – their genetic material is contained in a nucleus. They lack a rigid cell wall. Other structures (organelles) in the cell cytoplasm have specialized roles.

- **MITOCHONDRION**: Releases energy in the cell
- **SMOOTH ENDOPLASMIC RETICULUM**: Helps make fats and other vital compounds
- **NUCLEUS**: Contains the cell's DNA
- **ROUGH ENDOPLASMIC RETICULUM**: Involved in making proteins (see pp.92–93)
- **CENTRIOLE**: Part of the cell's "skeleton" that helps organize the cell contents
- **MEMBRANE**: Controls what enters and leaves the cell
- **CYTOPLASM**: A gelatinous fluid
- **GOLGI COMPLEX**: Processes and sorts large molecules
- **RIBOSOME**: A protein factory
- **LYSOSOME**: Contains enzymes to break down molecules

CELLS | 15

NUCLEUS

Chromosomes are separated from the cell cytoplasm by the nuclear membrane.

PAIRED CHROMOSOMES
The cell has two copies of every chromosome. Members of a pair are similar but not identical and are called homologues.

SEX CHROMOSOMES
In eukaryotes, a pair of non-identical chromosomes determines the sex of the organism (see p.19).

CHROMOSOME SHAPE
Chromosomes in a cell vary in size and shape.

CELL

THE INFORMATION CENTRE

The nucleus of almost every cell in the body of a eukaryotic organism (see pp.14–15) contains a copy of the genetic information about that organism. It is held in the form of long, ladderlike molecules of DNA (deoxyribonucleic acid). In eukaryotic cells, DNA is coiled and packaged within structures called chromosomes. The number of chromosomes within a nucleus depends on the species (see p.18), but chromosomes are mostly present in pairs – one inherited from each parent in sexual reproduction.

DNA
DNA is a long molecule that carries the genetic code in its chemical structure (see pp.84–93).

GENE
A gene is a section of DNA that instructs the cell to make a particular protein.

The double helix
DNA has a characteristic twisted double-helix shape. It is packaged within chromosomes.

16 | GENES AND CHROMOSOMES

FROM GENES TO PROTEINS

A gene is a section of DNA. It carries a set of coded instructions that instruct a cell to make a particular protein, and when to make it. Proteins are large, complex organic molecules that have critical roles in the body (see pp.85–87). Some are structural, while others control chemical processes that are necessary for life. Genes work by regulating the manufacture of diverse proteins within a cell through the processes of transcription and translation (see pp.90–93).

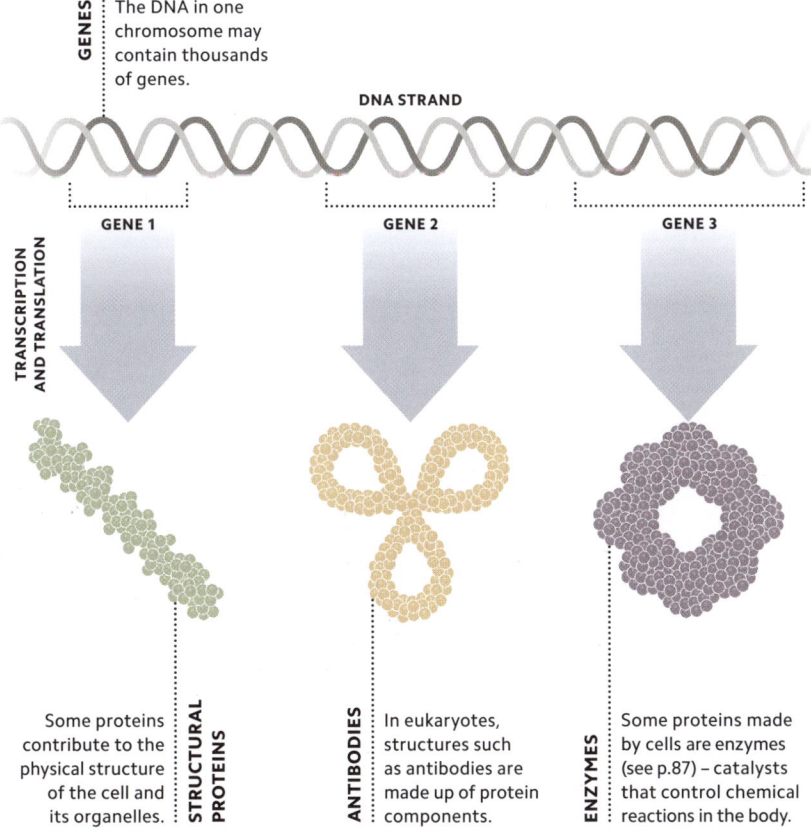

GENES The DNA in one chromosome may contain thousands of genes.

DNA STRAND

GENE 1 GENE 2 GENE 3

TRANSCRIPTION AND TRANSLATION

STRUCTURAL PROTEINS Some proteins contribute to the physical structure of the cell and its organelles.

ANTIBODIES In eukaryotes, structures such as antibodies are made up of protein components.

ENZYMES Some proteins made by cells are enzymes (see p.87) – catalysts that control chemical reactions in the body.

MORE OR LESS

All individuals of one species have the same number of chromosomes and a similar (but not identical) set of genes. Species differ greatly in the number of chromosomes and genes they possess. Most animals have two copies of each chromosome in every cell but some, such as male bees and ants, have only one set because they are the result of asexual reproduction (see p.22 and p.67). Some plants have multiple copies of each chromosome (see p.130).

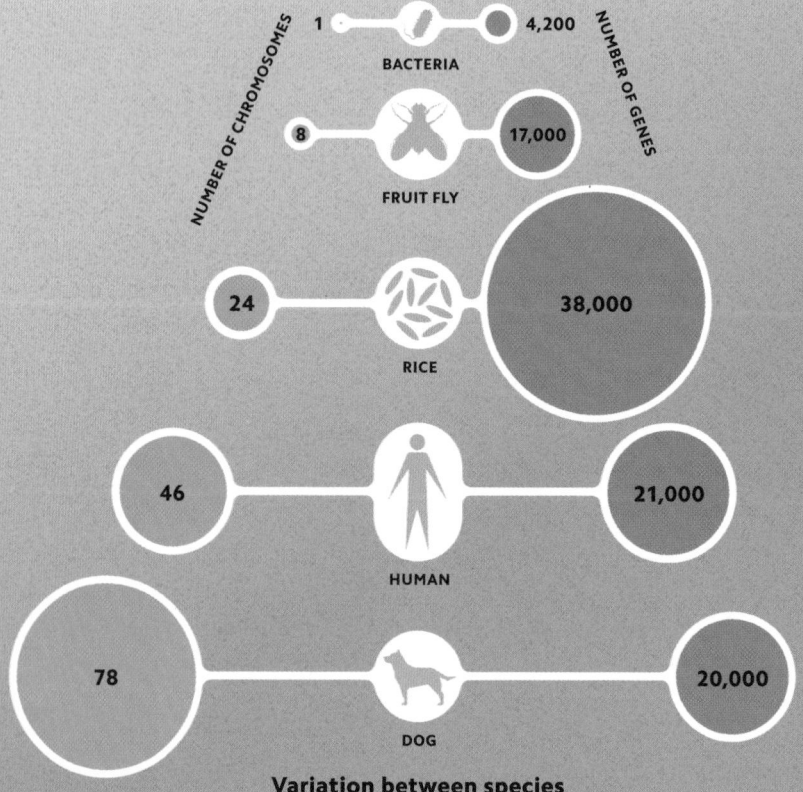

Variation between species
The number of chromosomes within a cell is not proportional to the number of genes.

CHROMOSOME NUMBERS

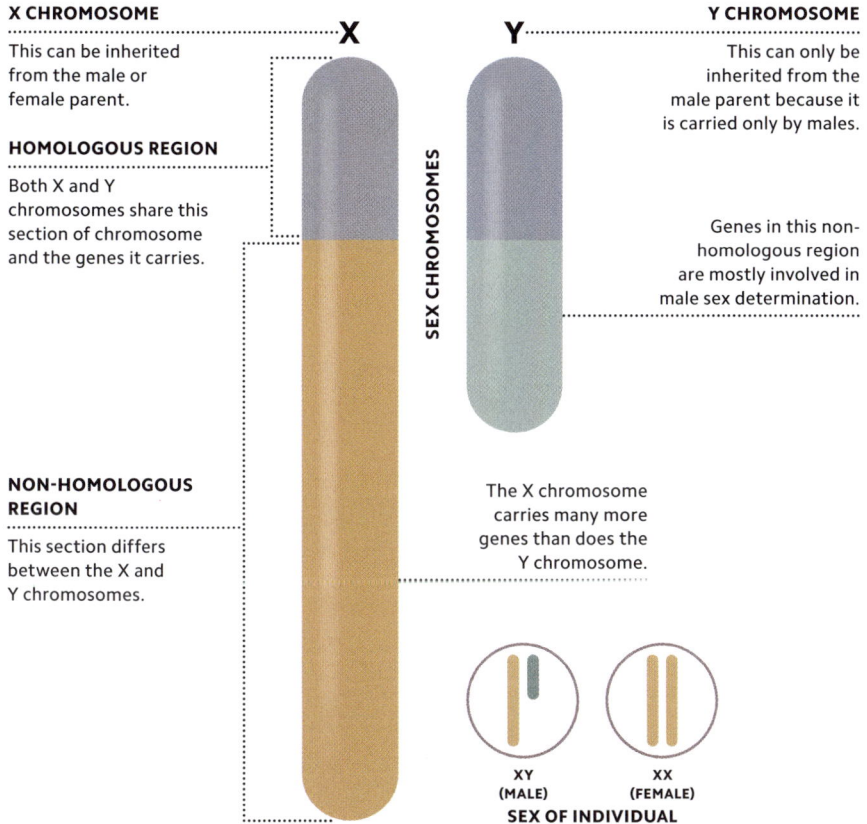

X CHROMOSOME
This can be inherited from the male or female parent.

HOMOLOGOUS REGION
Both X and Y chromosomes share this section of chromosome and the genes it carries.

NON-HOMOLOGOUS REGION
This section differs between the X and Y chromosomes.

Y CHROMOSOME
This can only be inherited from the male parent because it is carried only by males.

Genes in this non-homologous region are mostly involved in male sex determination.

The X chromosome carries many more genes than does the Y chromosome.

XY (MALE)   XX (FEMALE)
SEX OF INDIVIDUAL

A MATTER OF SEX

Most animals and plants have two copies of each chromosome – one inherited from each parent. Members of each chromosome pair are alike in terms of their size and shape, and they carry similar genes. However, the chromosomes that determine the sex of the organism may be a similar or different pair. In mammals, including humans, females have a pair of similar chromosomes, denoted XX, while the male sex chromosomes are dissimilar, denoted XY.

CHROMOSOME PATTERNS

Chromosomes can be stained and viewed under a microscope. Such observations reveal that species differ in chromosome number, but also chromosome lengths, shapes, and banding patterns. The image of a full set of chromosomes from an individual arranged in size order is called a karyotype. Early geneticists compared karyotypes of different species and observed similarities in related species.

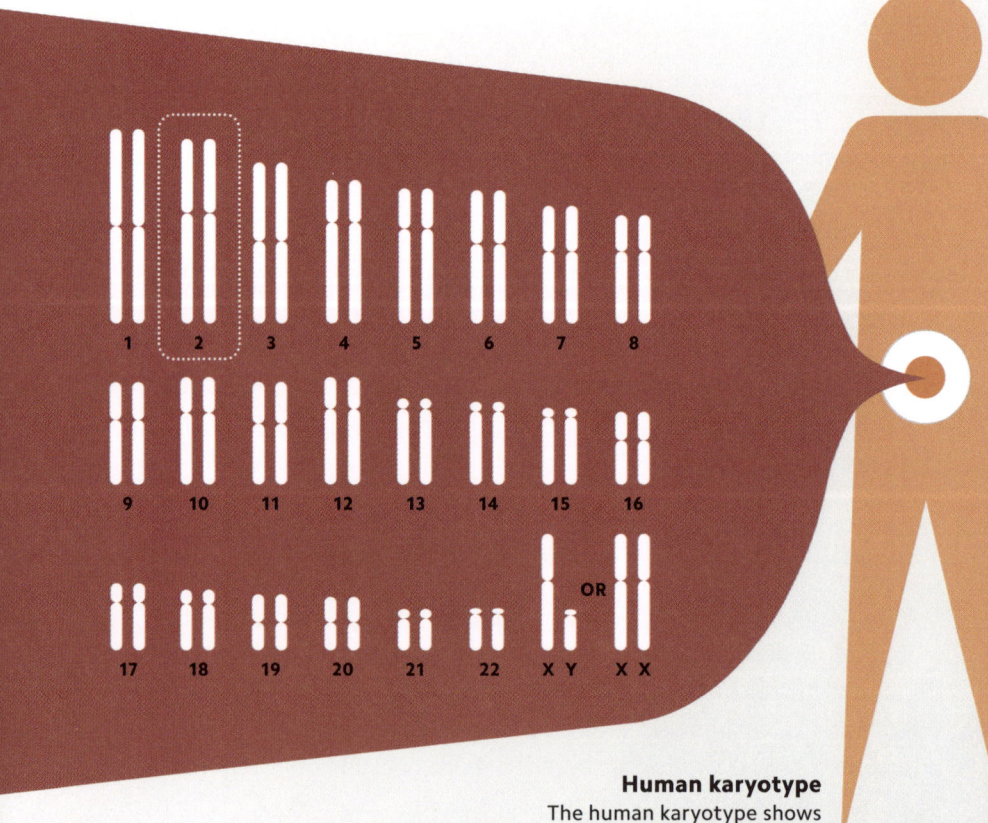

Human karyotype
The human karyotype shows 23 pairs of chromosomes.

Humans and chimpanzees evolved from a single common ancestor around six million years ago.

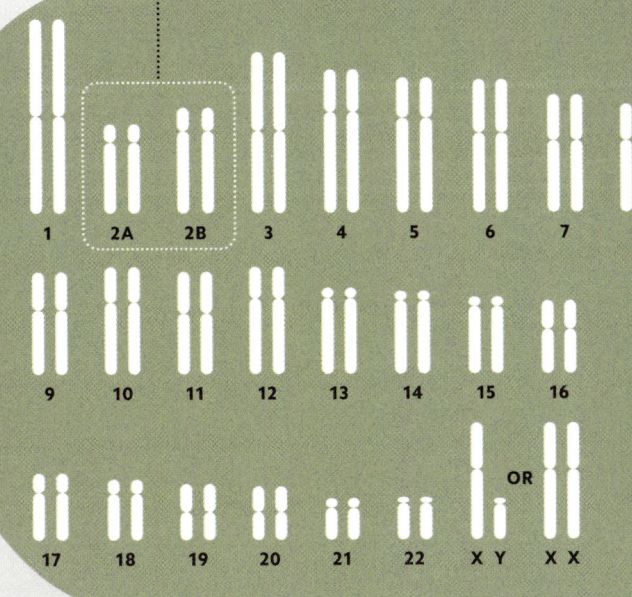

Chromosomes 2A and 2B fused in the evolution of humans but remained separate in chimpanzees.

Chimpanzee karyotype
Chimpanzees have 24 pairs of chromosomes, one more than humans.

KARYOTYPES | 21

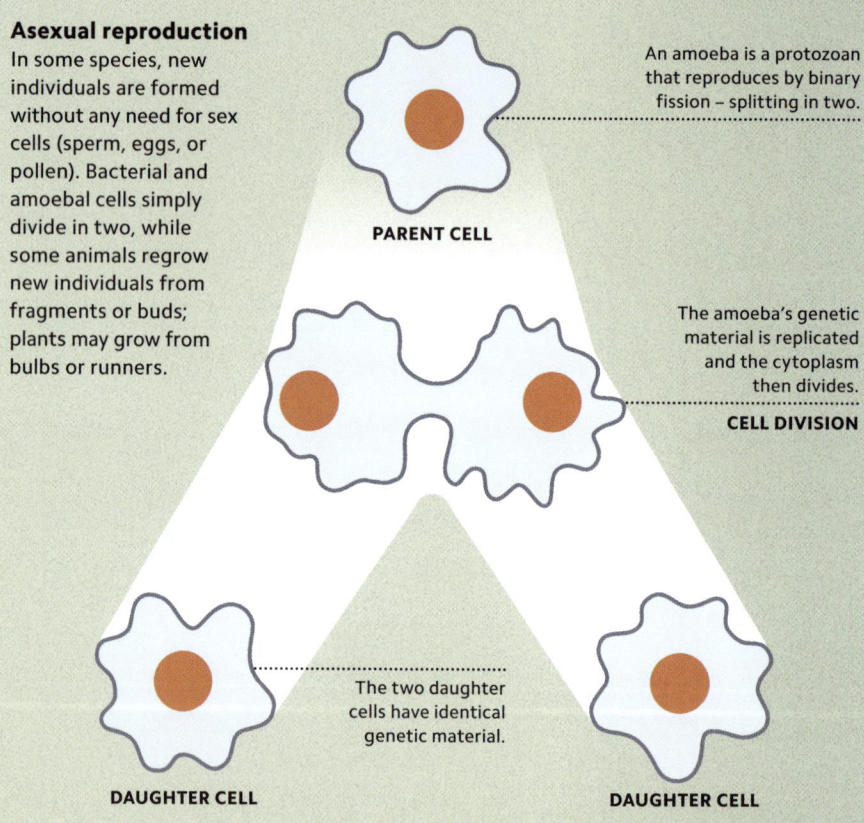

Asexual reproduction
In some species, new individuals are formed without any need for sex cells (sperm, eggs, or pollen). Bacterial and amoebal cells simply divide in two, while some animals regrow new individuals from fragments or buds; plants may grow from bulbs or runners.

An amoeba is a protozoan that reproduces by binary fission – splitting in two.

PARENT CELL

The amoeba's genetic material is replicated and the cytoplasm then divides.
CELL DIVISION

The two daughter cells have identical genetic material.

DAUGHTER CELL **DAUGHTER CELL**

THE CONTINUITY OF LIFE

Life does not form from nothing. The members of one generation grow, develop, and reproduce, creating the next. Living things are made from cells, and it is the division of these cells that produces new life. Some organisms arise when cells divide directly to form identical offspring. This is called asexual reproduction. Most organisms arise from sexual reproduction, which involves the production and fusion of specialized sex cells, or gametes.

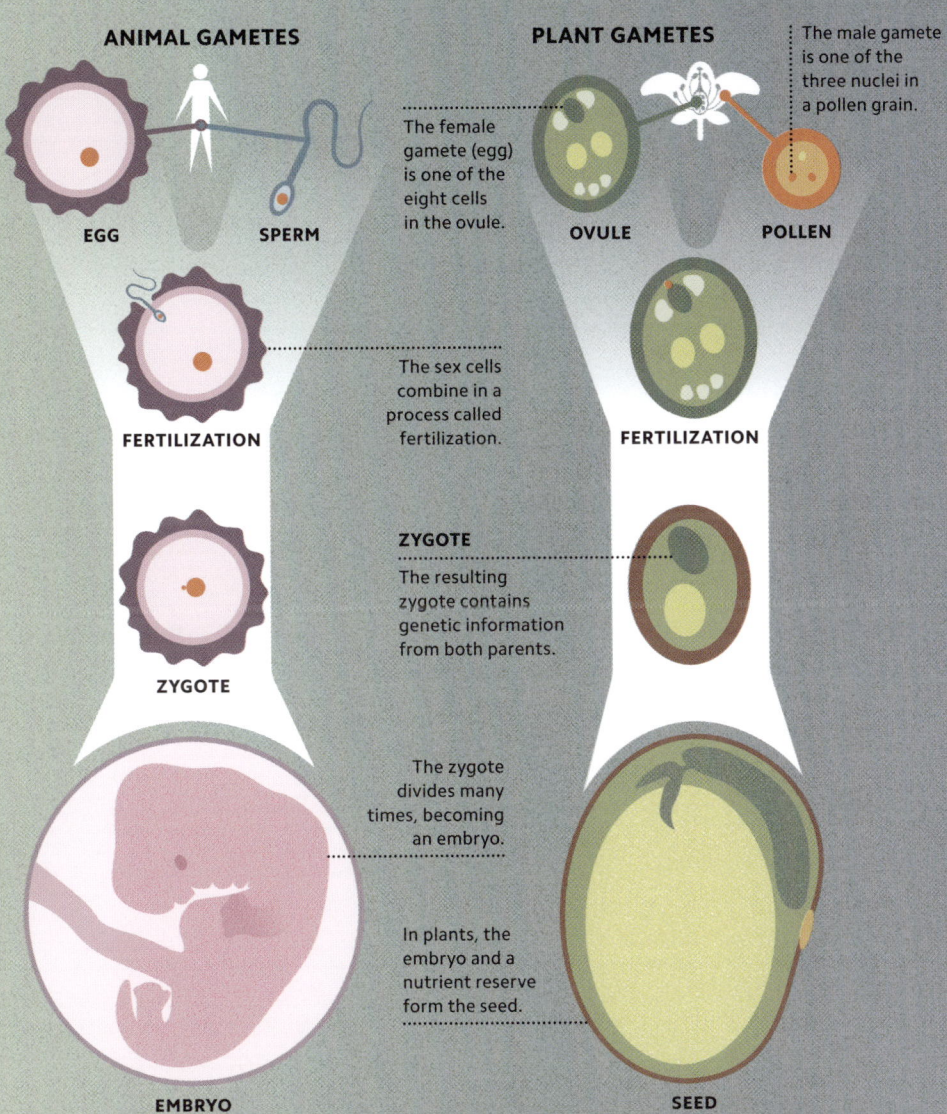

Sexual reproduction
Most plants and animals reproduce by sexual reproduction, even though this involves investing in the production of specialized sex cells (see p.24). Its advantage is that it generates variability in the offspring (see pp.78–81).

TYPES OF REPRODUCTION | 23

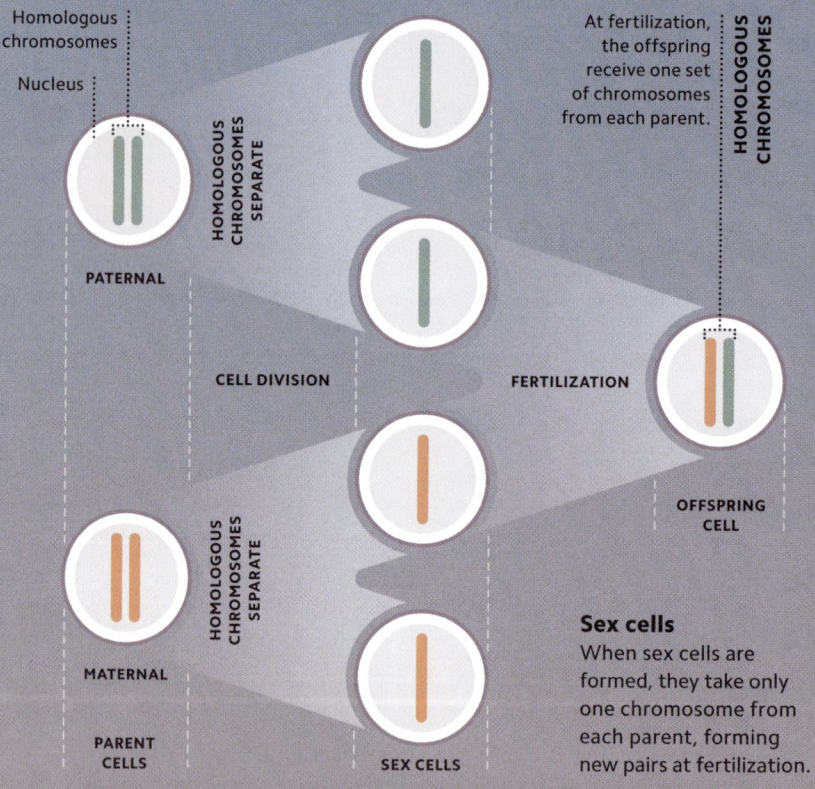

Sex cells
When sex cells are formed, they take only one chromosome from each parent, forming new pairs at fertilization.

HALF AND HALF

Sexually reproducing organisms inherit a set of genes from each parent. They possess two copies of each chromosome – one of maternal and one of paternal origin. These chromosome pairs, called homologous pairs, have the same features – they are the same size and shape, and have the same genes at the same positions along their length (with the exception of the sex chromosomes; see p.19). Homologous chromosomes must be separated when sex cells are formed, otherwise chromosome numbers would double with each generation.

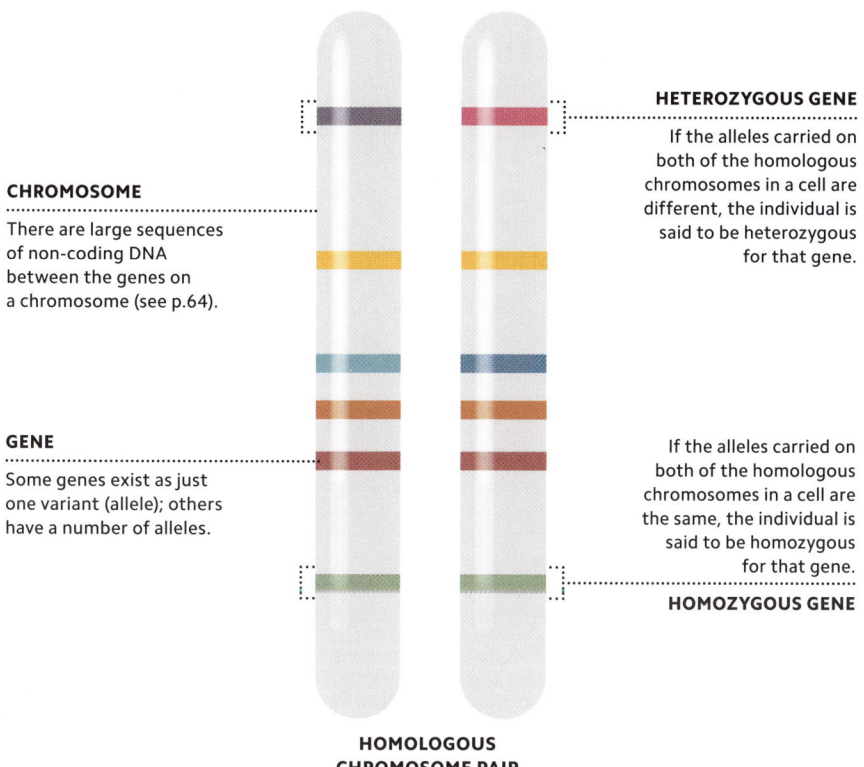

CHROMOSOME
There are large sequences of non-coding DNA between the genes on a chromosome (see p.64).

GENE
Some genes exist as just one variant (allele); others have a number of alleles.

HETEROZYGOUS GENE
If the alleles carried on both of the homologous chromosomes in a cell are different, the individual is said to be heterozygous for that gene.

If the alleles carried on both of the homologous chromosomes in a cell are the same, the individual is said to be homozygous for that gene.
HOMOZYGOUS GENE

HOMOLOGOUS CHROMOSOME PAIR

VARIANTS OF GENES

Some genes occur in a variety of forms, called alleles. Pea plants, for example, have a gene that determines the colour of their flowers, which can be white or purple. This is because there are two variants of the flower-colour gene – white and purple (see pp.36–37). The plant cells have two copies of every chromosome, so carry two copies of every gene. If the alleles on both chromosomes are the same, the plant is said to be homozygous for that gene; if the alleles are different, it is heterozygous.

ALLELE INTERACTIONS

Sexually reproducing organisms have two copies of each gene. These copies, called alleles, can be slightly different from each other and so produce slightly different characteristics. If two different alleles are present in one individual, one allele may overrule the effects of the other. The effects of this so-called dominant allele are evident, while the effects of the other – the recessive allele – remain hidden.

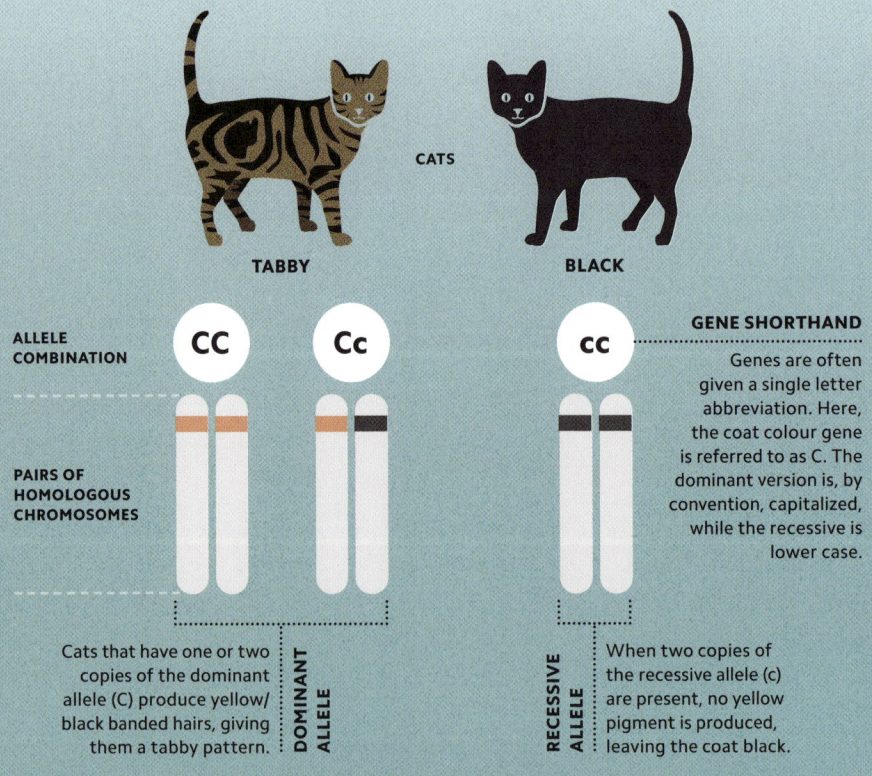

CATS

TABBY — BLACK

ALLELE COMBINATION: CC — Cc — cc

PAIRS OF HOMOLOGOUS CHROMOSOMES

GENE SHORTHAND
Genes are often given a single letter abbreviation. Here, the coat colour gene is referred to as C. The dominant version is, by convention, capitalized, while the recessive is lower case.

DOMINANT ALLELE: Cats that have one or two copies of the dominant allele (C) produce yellow/black banded hairs, giving them a tabby pattern.

RECESSIVE ALLELE: When two copies of the recessive allele (c) are present, no yellow pigment is produced, leaving the coat black.

Cat coat colour
The presence of just one dominant allele (C) confers tabby coat colour in cats. Black colour demands two recessive alleles (c).

DOMINANT OR RECESSIVE

> Dominant genes are not always "better" than recessive ones, some of which confer a benefit.

CATTLE

| PURE RED | ROAN | WHITE |

RR — RED DOMINANT
An animal that carries two copies of the R gene has a red coat.

RW — CO-DOMINANT
The presence of both dominant alleles results in a roan coat.

WW — WHITE DOMINANT
An animal that carries two copies of the W gene is white.

Roan co-dominance
Sometimes, two or more alleles can carry equal weight, in which case they are called co-dominant. The roan coat gene in horses and cattle works in this way.

DOMINANT OR RECESSIVE | 27

INVESTI
INHERIT

The systematic study of genetics began when naturalists attempted to unravel the laws of inheritance. They did this by carrying out breeding experiments using plant and animal species that had clear-cut and easily recognized varieties. The pioneer geneticists crossbred varieties, counted offspring, and looked for patterns in their numbers. The studies conducted in the 19th century by Gregor Mendel (with pea plants) and in the early 20th century by Thomas Morgan (with fruit flies) helped to develop ideas about inheritance that still form the basis for the study of genetics today.

NESTED LIFE?

A microscopic fertilized egg develops into the highly complex body of an animal. Some early biologists accounted for this fact by proposing that sex cells (sperm or eggs) contained tiny preformed bodies that grew bigger after birth and that each generation contained the next. We now know that life's continuity comes not from generations nested within generations, but from a remarkable material – today recognized as DNA – that governs development.

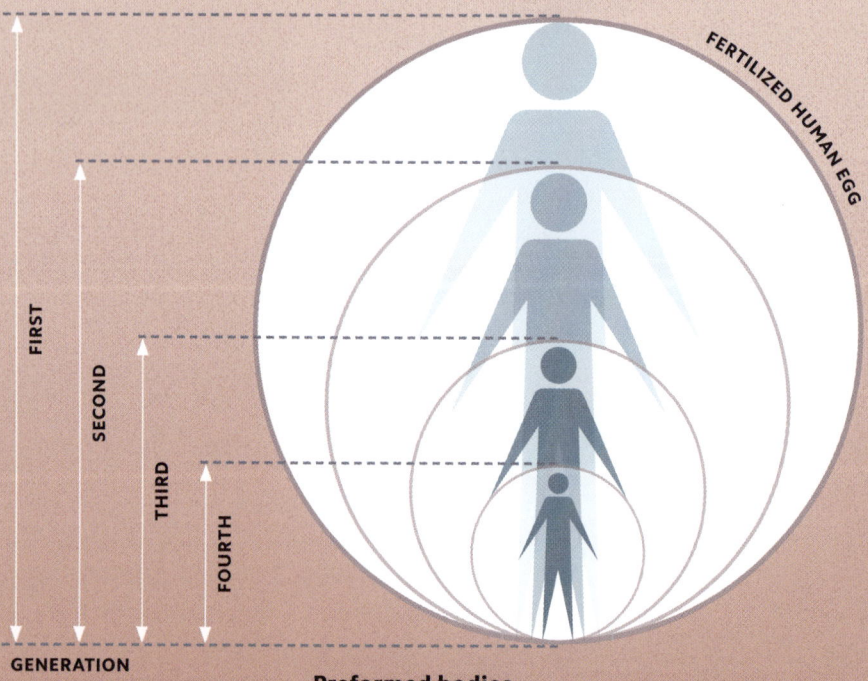

Preformed bodies
The theory of preformation held that tiny preformed individuals were present in sex cells, so one generation already contained the next. This idea avoided the tricky question of how structured bodies could be formed from less-structured matter.

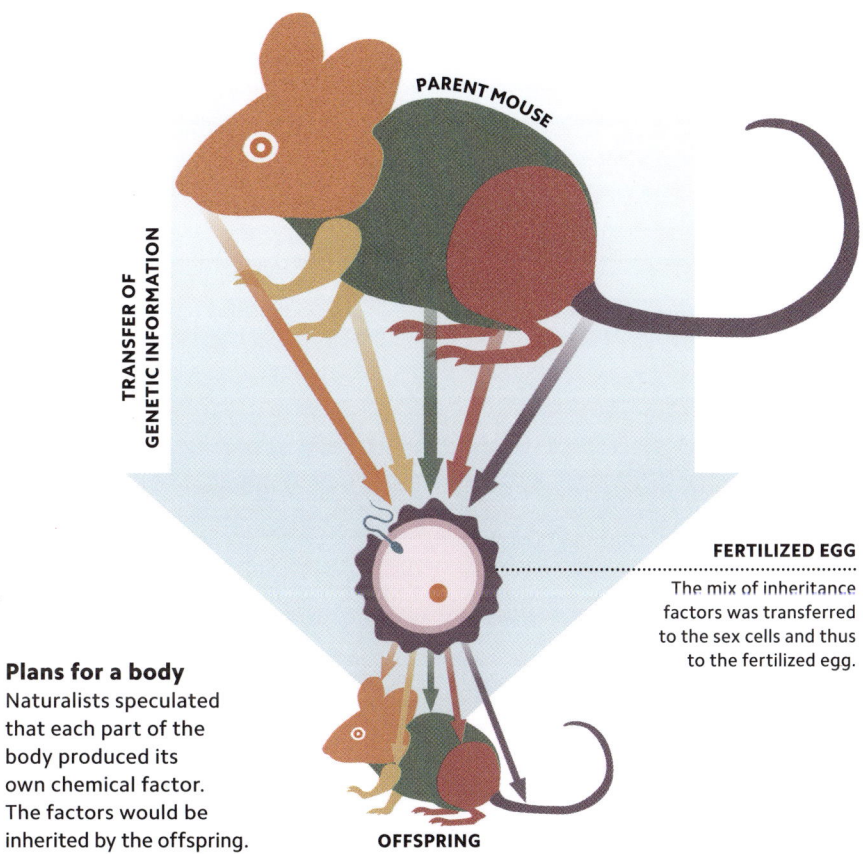

Plans for a body
Naturalists speculated that each part of the body produced its own chemical factor. The factors would be inherited by the offspring.

FERTILIZED EGG
The mix of inheritance factors was transferred to the sex cells and thus to the fertilized egg.

GENETIC FACTORS?

Naturalists attempted to account for inheritance by proposing that chemical "factors" produced by the body dictated how a fertilized egg developed into an adult. Each part of the body would produce its own factor that was transmitted through the sex cells. However, experiments showed that animals with missing body parts would still produce intact offspring. This suggested that the full complement of development factors was present in every part of the anatomy.

LIFE IN THE MIX

BLENDED COLOURS
All first-generation offspring had blended, pink-coloured flowers.

FOUR O'CLOCK PLANT

FIRST CROSS
Red and white varieties were crossed. Each came from a long line of red-only or white-only plants.

RED FLOWER

WHITE FLOWER

PARENTAL VARIETIES

FIRST-GENERATION OFFSPRING

Early biologists speculated that the "factors" (see p.31) governing the development of body parts were blended in sexual reproduction, much as red paint can be blended with white to make pink. According to this idea, variations would eventually merge into one form, smoothing out differences over the generations. However, the results of breeding experiments showed something surprising – that characteristics could pass through generations unchanged, in their original form.

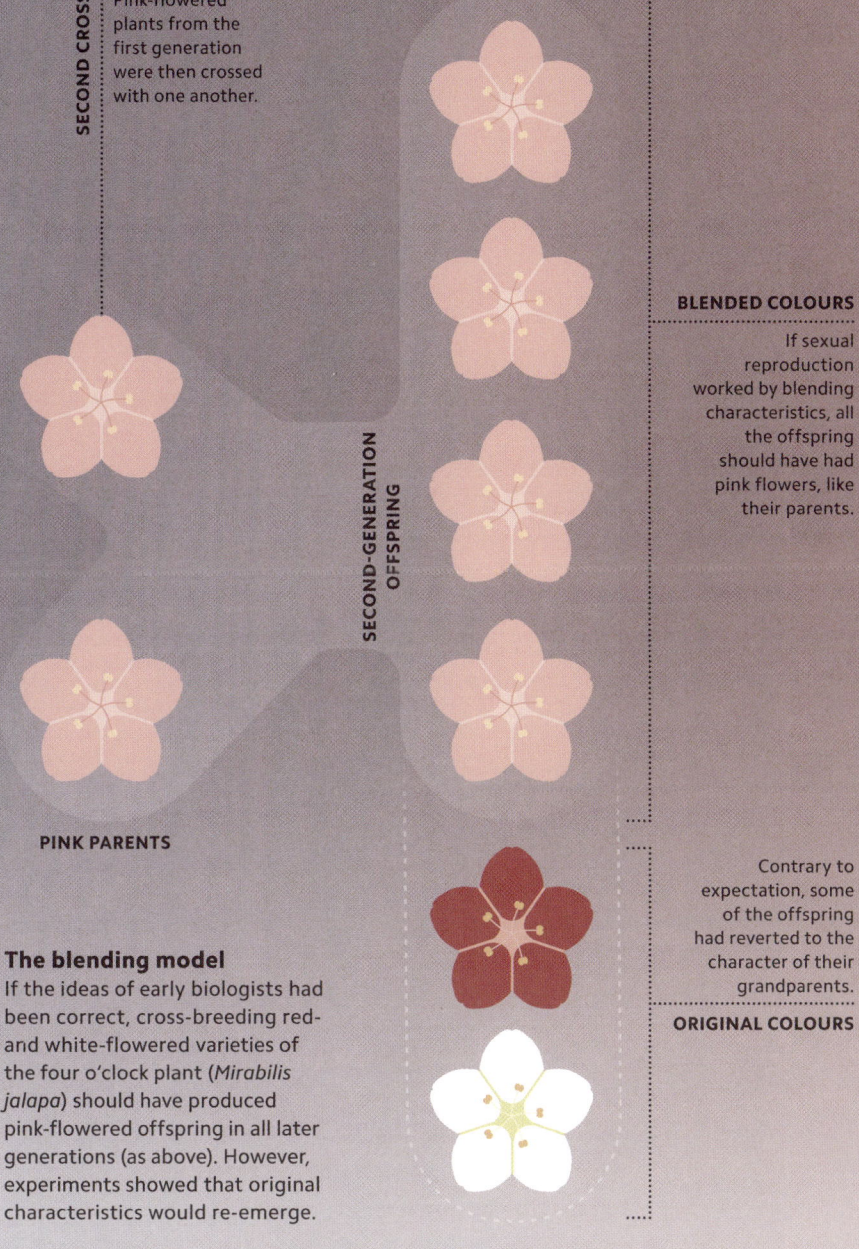

SECOND CROSS — Pink-flowered plants from the first generation were then crossed with one another.

SECOND-GENERATION OFFSPRING

BLENDED COLOURS — If sexual reproduction worked by blending characteristics, all the offspring should have had pink flowers, like their parents.

PINK PARENTS

The blending model
If the ideas of early biologists had been correct, cross-breeding red- and white-flowered varieties of the four o'clock plant (*Mirabilis jalapa*) should have produced pink-flowered offspring in all later generations (as above). However, experiments showed that original characteristics would re-emerge.

Contrary to expectation, some of the offspring had reverted to the character of their grandparents.

ORIGINAL COLOURS

MECHANICS OF INHERITANCE | 33

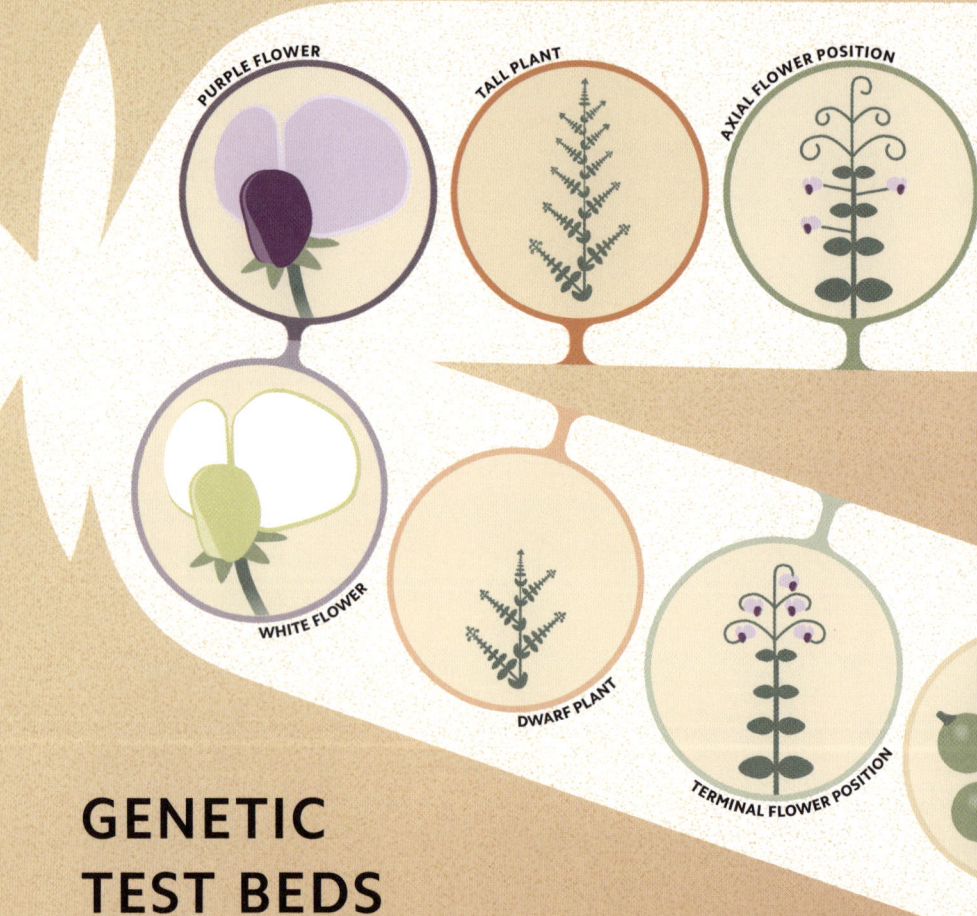

GENETIC TEST BEDS

The best organisms to use in breeding experiments are ones that have short lifespans and reproduce prolifically. The more offspring, the more data is available to reveal statistically significant patterns of inheritance. In the 1850s, Austrian monk Gregor Mendel carried out breeding studies on garden peas; other organisms popular with geneticists are fruit flies, yeasts, nematode worms, and mice. Mendel's experiments were groundbreaking because he explored the inheritance of individual characteristics, analysing them mathematically, in contrast to the more holistic approach of many of his contemporaries.

DOMINANT CHARACTERISTICS

VARIATIONS

> Mendel is known as the "father of genetics" for his groundbreaking studies.

Pea studies
Garden peas are useful test beds for genetic studies because they have a number of characteristics that clearly vary between individuals, so the outcomes of different crosses can be assessed easily and produce plenty of data.

MODEL ORGANISMS | 35

3:1 INHERITANCE

In the 1850s, Austrian monk Gregor Mendel carried out breeding experiments using pea plants. He crossed varieties with purple and white flowers and counted the numbers of each flower colour in the offspring. He found that the flowers were all purple in the first generation of offspring. However, white flowers reappeared in the next generation, accounting for one-quarter of the offspring. To Mendel, this mathematical ratio was very significant and pointed to how the "factors" of inheritance behaved.

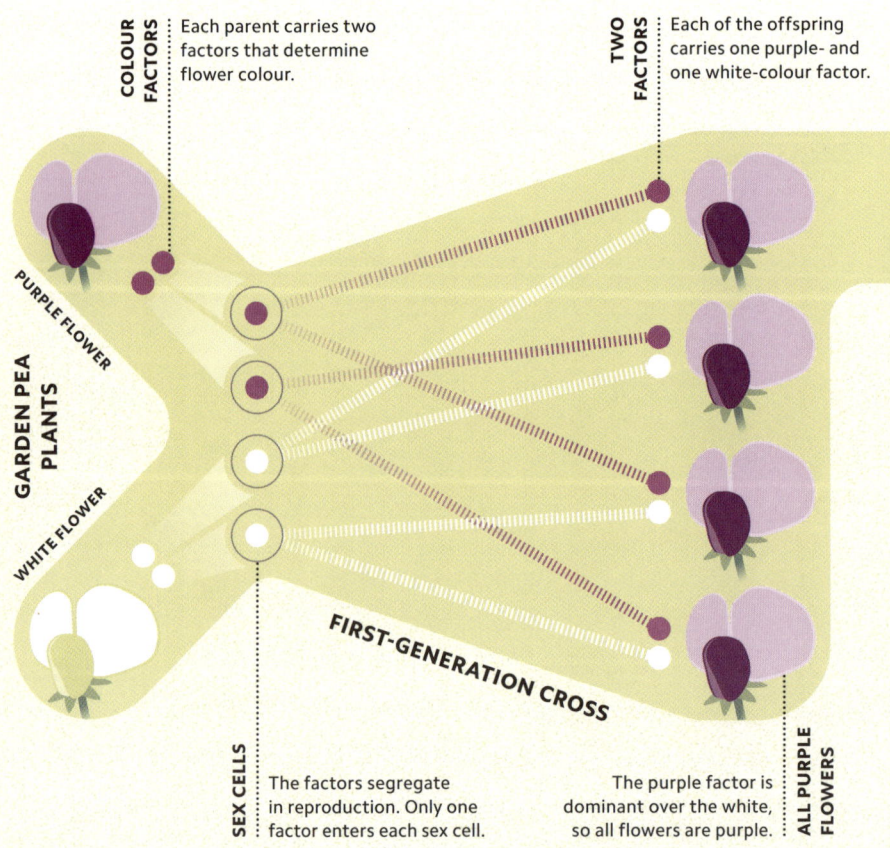

COLOUR FACTORS: Each parent carries two factors that determine flower colour.

TWO FACTORS: Each of the offspring carries one purple- and one white-colour factor.

GARDEN PEA PLANTS — PURPLE FLOWER / WHITE FLOWER

FIRST-GENERATION CROSS

SEX CELLS: The factors segregate in reproduction. Only one factor enters each sex cell.

The purple factor is dominant over the white, so all flowers are purple.

ALL PURPLE FLOWERS

36 | SIMPLE CROSSES

Segregation of factors

The ratios revealed by Mendel's experiments could only be explained if two factors of inheritance governed each characteristic. These factors are carried in pairs by the parent plants. One type of factor (purple flowers) masks, or is dominant over, the other, which is termed "recessive". The two factors within the parents move into separate sex cells (pollen or eggs) during reproduction.

> Mendel published his theory of inheritance in 1866; it was another 35 years before it was rediscovered and accepted.

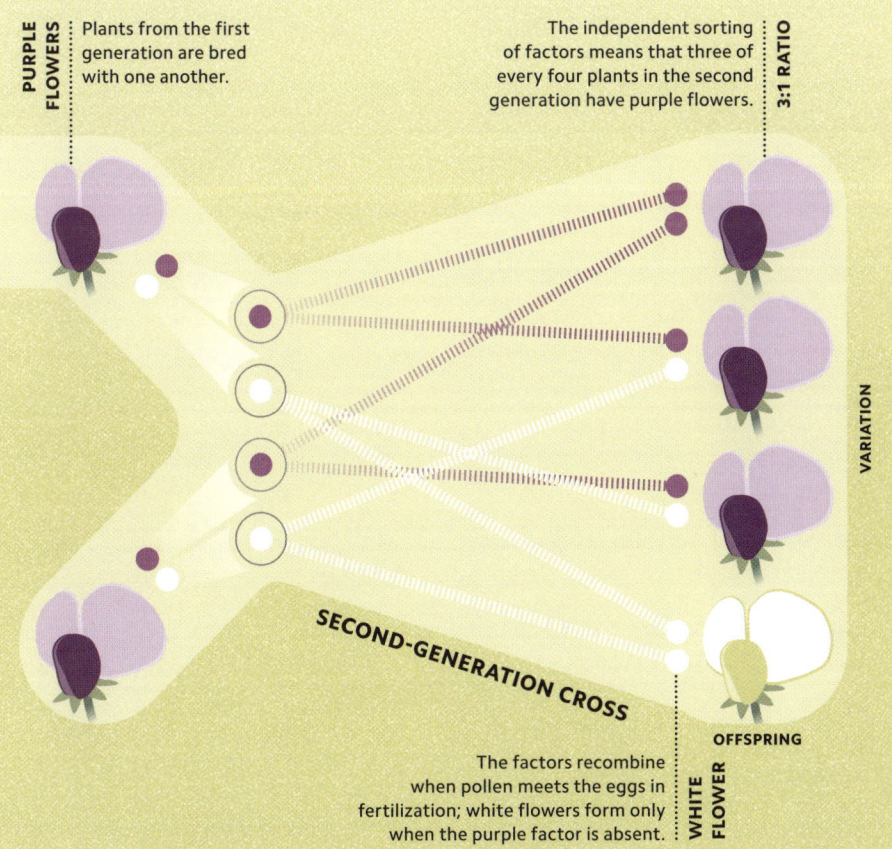

PURPLE FLOWERS

Plants from the first generation are bred with one another.

The independent sorting of factors means that three of every four plants in the second generation have purple flowers.

3:1 RATIO

VARIATION

SECOND-GENERATION CROSS

The factors recombine when pollen meets the eggs in fertilization; white flowers form only when the purple factor is absent.

WHITE FLOWER

OFFSPRING

SIMPLE CROSSES | 37

RATIOS OF INHERITANCE

Mendel revealed the basic patterns of heredity by tracking just one characteristic – flower colour – in his first breeding experiments (see pp.36–37). He went on to investigate the inheritance of two characteristics at the same time. These so-called dihybrid crosses considered the inheritance of purple/white flowers and green/yellow pea pods. Once again, the dominant factors (purple flowers and green pods) masked the white-flower and yellow-pod elements in all the first-generation offspring. When crossing plants from that generation, every possible character combination came through in the next generation; the ratios of characteristics in these offspring confirmed Mendel's proposed mechanism of inheritance.

Independent factors
Mendel predicted that the result of two characters being inherited independently would yield a specific ratio of offspring type – 9:3:3:1. His breeding experiments – the results of which are shown here – confirmed his mathematical prediction.

A purple-flowered plant that produces green pods is crossed with a white-flowered plant that produces yellow pods.

PARENT PLANTS

The sex cells (pollen and eggs) carry one factor for flower colour and one for pod colour.

SEX CELLS

GARDEN PEA PLANTS

PUREBRED PARENTS

OFFSPRING

FIRST-GENERATION CROSS

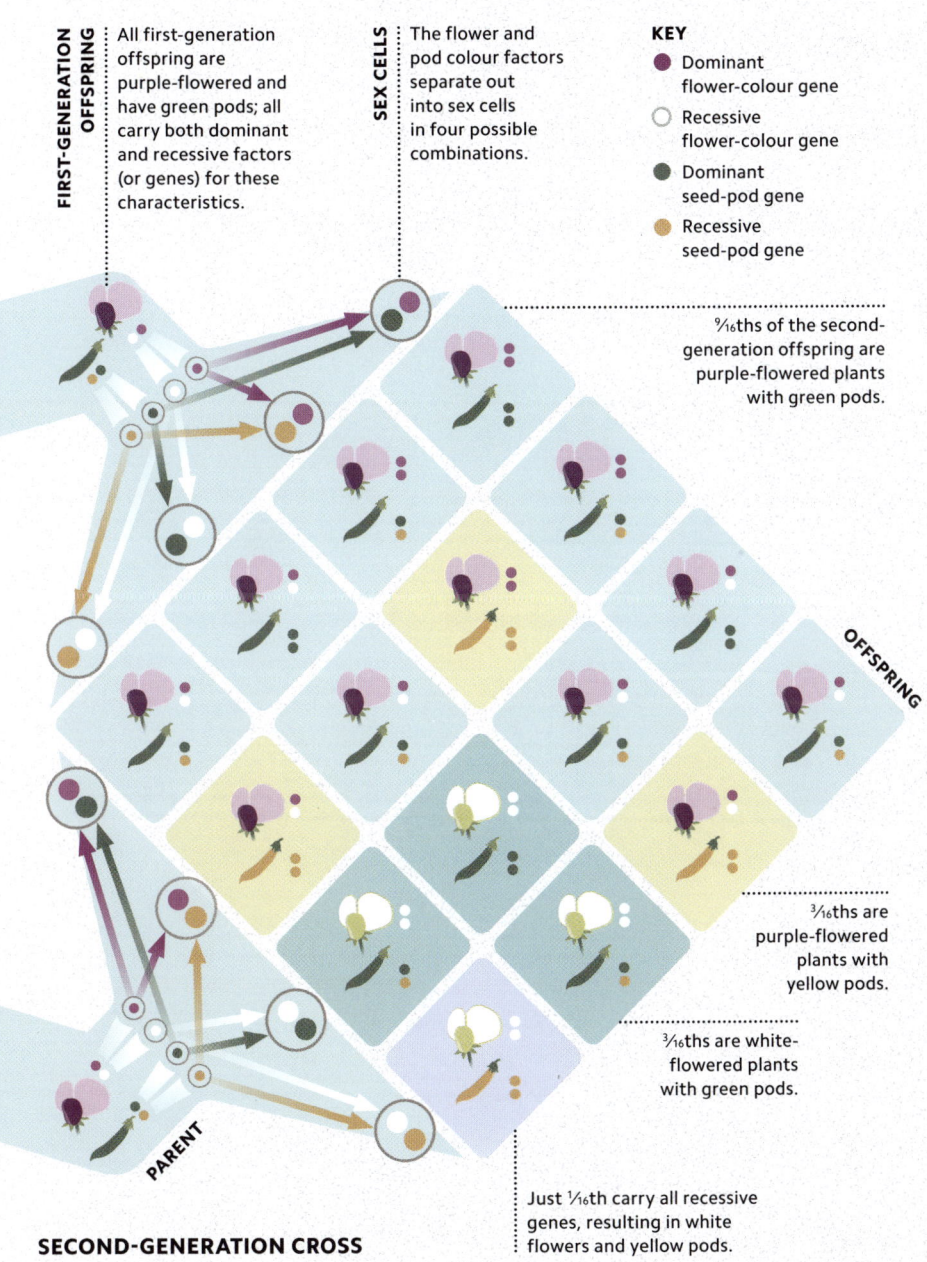

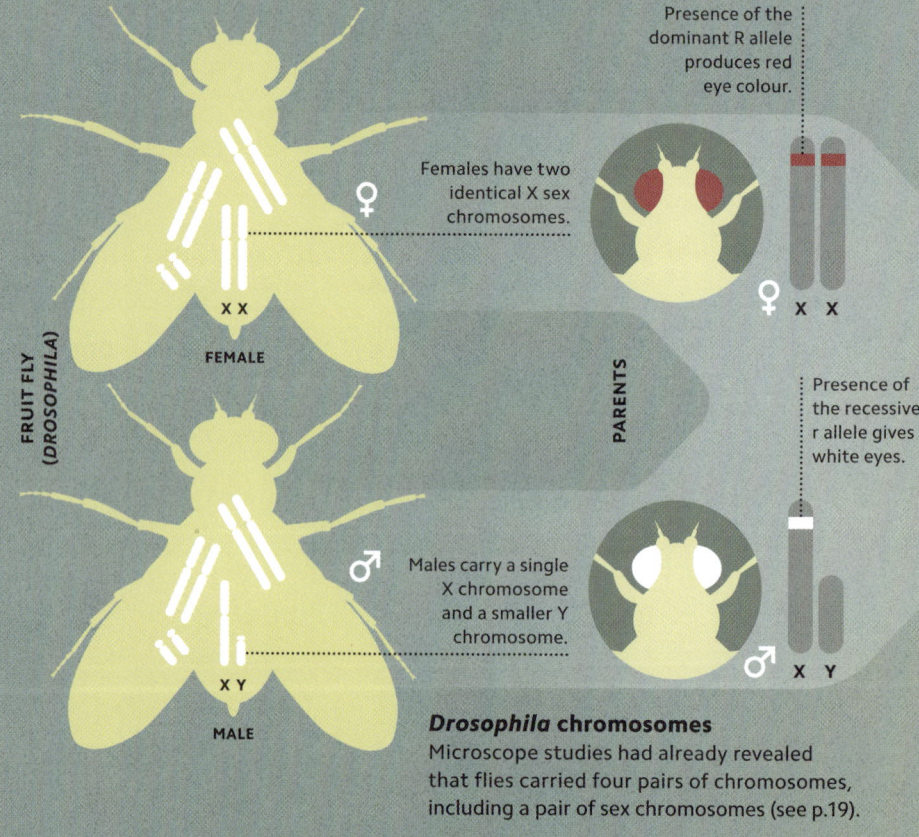

Drosophila chromosomes
Microscope studies had already revealed that flies carried four pairs of chromosomes, including a pair of sex chromosomes (see p.19).

FINDING THE FACTORS

In the early 20th century, biologists began to look for the physical location of Mendel's "factors" (see pp.36–39) in the cell. Geneticist Thomas Morgan, who was carrying out breeding studies on the fruit fly (*Drosophila melanogaster*), noticed that some male flies had mutant white (instead of normal red) eyes, but that females had white eyes only very rarely. He speculated that the white factor was carried on the fly's sex chromosome, so linking a factor with a physical location.

Morgan's experiments

Normal flies have red eyes, while mutant flies have white eyes. Morgan crossed a red-eyed female with a white-eyed male.

OFFSPRING

FIRST-GENERATION OFFSPRING

All offspring in the first generation had red eyes because the red allele is dominant over white. This was as expected following Mendel's principles of inheritance.

SECOND-GENERATION OFFSPRING

The second-generation cross produced red-eyed males and females, white-eyed males, but no white-eyed females. This deviated from Mendel's law of independent assortment, suggesting that eye colour was located on the X chromosome.

THE CHROMOSOME THEORY OF INHERITANCE

TIED TOGETHER

Early in the 20th century, Mendel's "factors" of inheritance came to be called genes (see p.7). Breeding experiments with fruit flies showed that genes were located on chromosomes (see pp.40–41). It followed that two genes on the same chromosome would be linked and so inherited together, as one unit. Thomas Morgan showed that this was the case, but that the degree of linkage was variable. This is because alleles could be swapped between homologous chromosomes during the formation of sex cells. This process is called crossing over (see p.78). Moreover, the further apart the alleles were on the chromosome, the more frequently they were swapped between homologues.

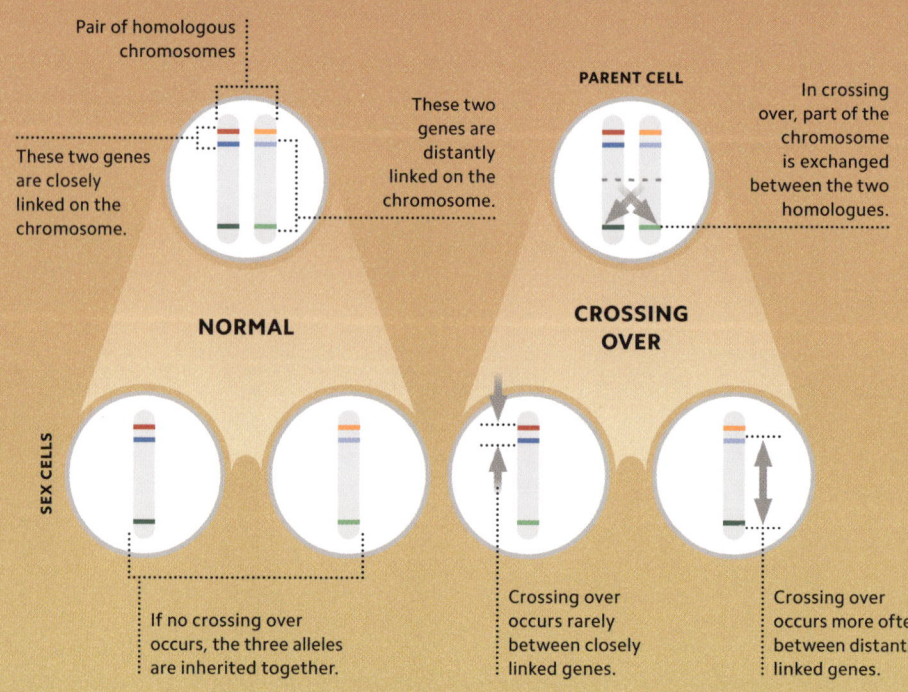

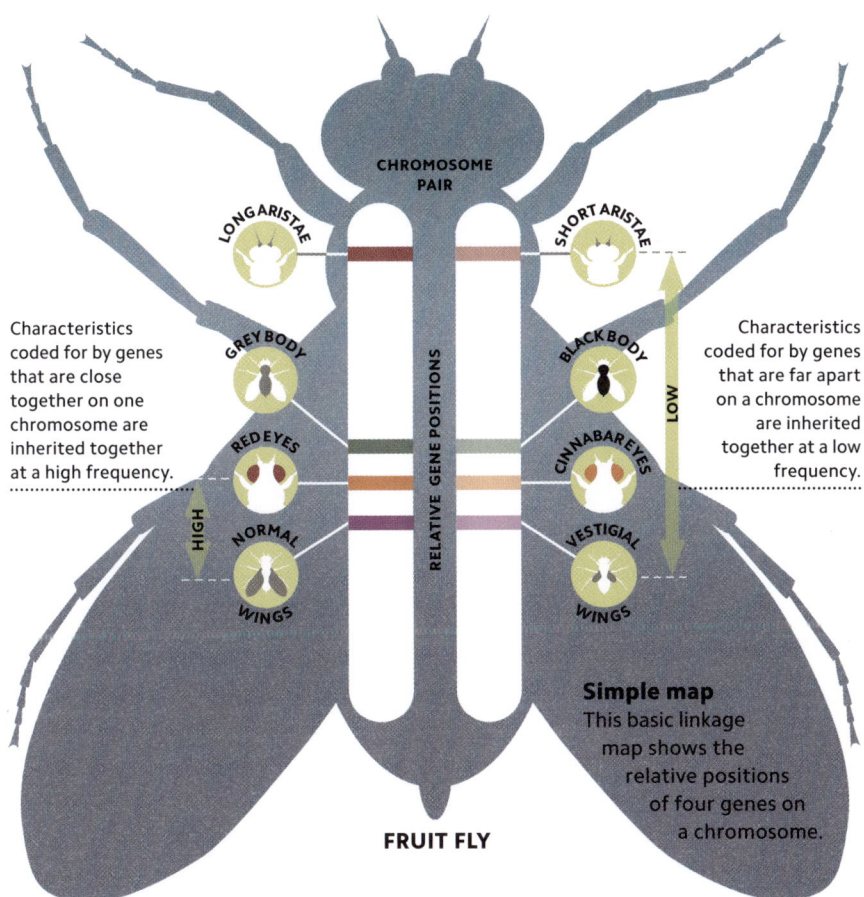

Simple map
This basic linkage map shows the relative positions of four genes on a chromosome.

GENE GEOGRAPHY

The statistical frequency with which two genes are inherited together is a measure of how close those genes are on a chromosome. By carrying out many thousands of crosses and comparing how often different pairs of genes are inherited together as a unit, geneticists can produce "linkage maps" that show the relative positions of all of the genes on a chromosome. The fruit fly (which breeds very quickly, producing a new generation every 10–12 days) proved to be the ideal subject for early studies of gene linkage (see pp.34–35).

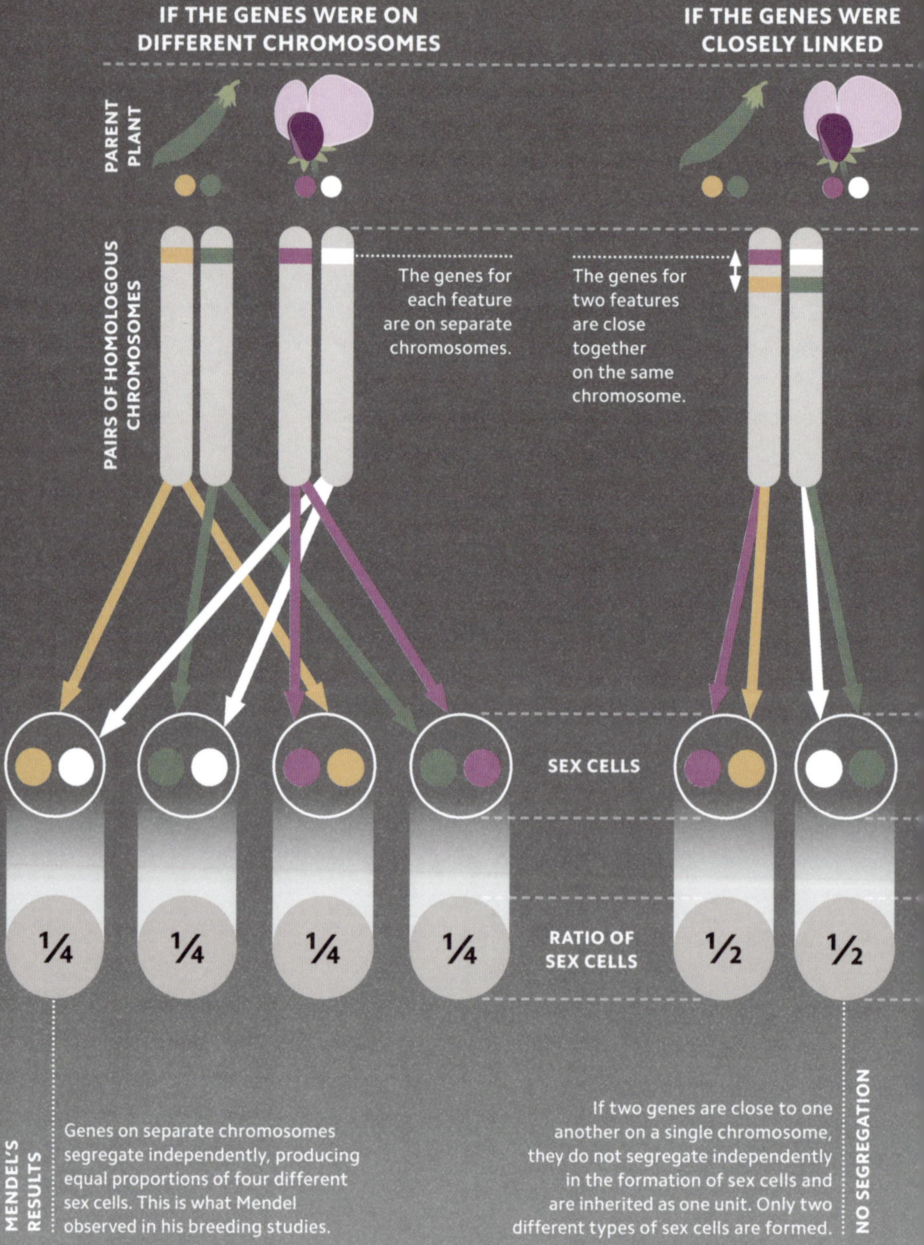

IF THE GENES WERE ON DIFFERENT CHROMOSOMES

IF THE GENES WERE CLOSELY LINKED

PARENT PLANT

PAIRS OF HOMOLOGOUS CHROMOSOMES

The genes for each feature are on separate chromosomes.

The genes for two features are close together on the same chromosome.

SEX CELLS

RATIO OF SEX CELLS

¼ ¼ ¼ ¼ ½ ½

MENDEL'S RESULTS

Genes on separate chromosomes segregate independently, producing equal proportions of four different sex cells. This is what Mendel observed in his breeding studies.

NO SEGREGATION

If two genes are close to one another on a single chromosome, they do not segregate independently in the formation of sex cells and are inherited as one unit. Only two different types of sex cells are formed.

44 | THE PHYSICAL BASIS OF INHERITANCE

IF THE GENES WERE DISTANTLY LINKED

The genes for two features are widely separated on the same chromosome.

Sections of chromosomes can be "swapped" between homologues. This process is unpredictable and is called "crossing over".

CROSSING OVER

Crossing over is a random event, though it is more likely between genes separated by a large distance. The proportions of sex cells formed cannot be accurately predicted.

MENDEL'S FORTUNE

Mendel's experiments with pea plants led him to a description of the principles of inheritance (see pp.36–39). However, he did not know what the heritable "factors" were made of or where they were physically located. Later discoveries recognized the factors as genes, and located them on chromosomes. We can now see that Mendel was lucky in his choice of factors (genes) to study, because they were on different chromosomes or so far apart that they segregated independently. If the genes had been linked, he would have received very different results, as the results of this thought experiment show.

THE PHYSICAL BASIS OF INHERITANCE

VARIA

TION

The millions of living species on our planet are a striking expression of genetic diversification, but genetic variation visibly exists within species, too. While some characteristics may be shaped directly by the environment, all the inherited attributes of a single individual come from thousands of genes within the cells of its body. Genes vary between individuals and mix during reproduction, and new gene variants (or alleles) can be formed by mutation. These are the processes that generate genetic variation, explain our individuality, and ultimately account for the extraordinary biological diversity of our planet.

GENOTYPE AND PHENOTYPE

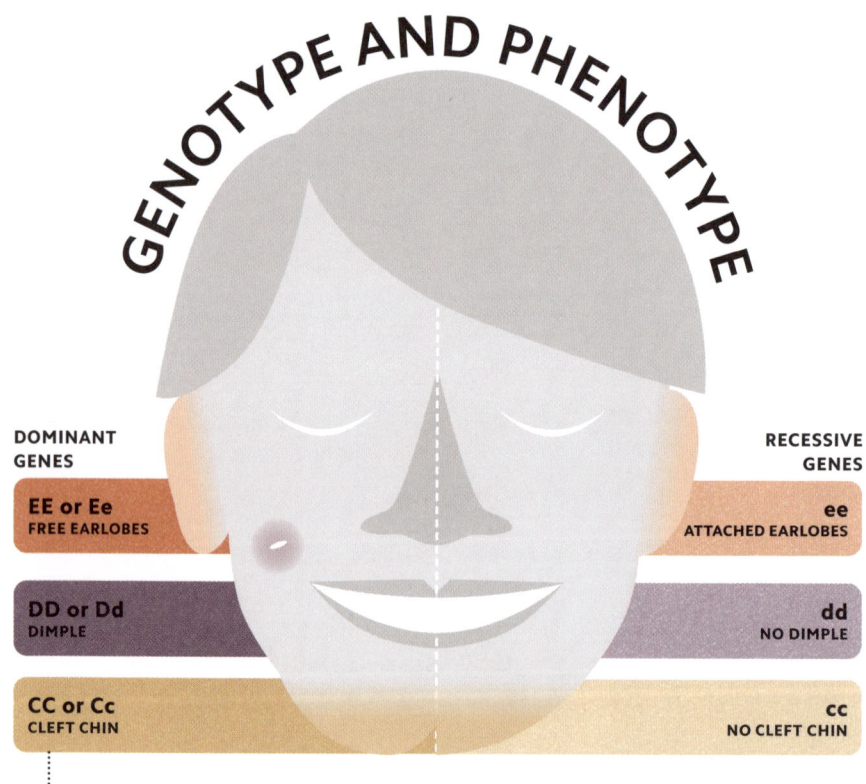

DOMINANT GENES
- EE or Ee — FREE EARLOBES
- DD or Dd — DIMPLE
- CC or Cc — CLEFT CHIN

RECESSIVE GENES
- ee — ATTACHED EARLOBES
- dd — NO DIMPLE
- cc — NO CLEFT CHIN

Genotype (here denoted by two letters) determines phenotype (observed characteristics).

Single gene control
Variability in some human facial features is controlled by a single gene.

Although organisms have thousands of different genes, studies into inheritance are usually confined to one or a few genes at a time, for practical reasons – it is very difficult to study the effects of many genes at the same time. Geneticists use the word "genotype" to describe a set of genes, distinguishing this from the "phenotype" – the observable physical properties of an organism, such as its appearance and behaviour. Genotype notation typically uses letters – upper and lower case, such as A and a – to indicate dominant and recessive alternatives.

STEPPED STATES

Some characteristics vary between one, two, or more distinct states – the flowers of pea plants, for example, are either purple or white, not any colour in between, and the antigens that determine human blood type may be A, B, or absent (O). Such discrete variation suggests a correspondence between one gene variant (allele) and one characteristic. In other cases, more than one gene may be involved in regulating one characteristic (see pp.50–52), or genes may interact with other genes to give more complex patterns of variation.

Blood types
Human blood type may be A, B, AB, or O. It is controlled by three alleles – A, B, and O; O is a recessive allele.

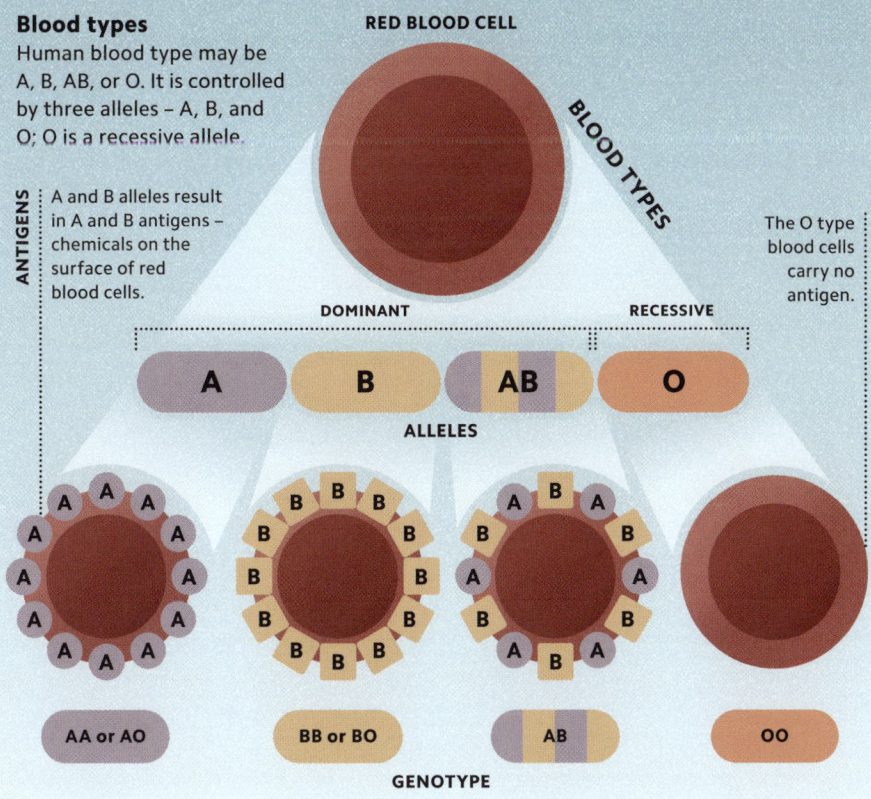

ANTIGENS: A and B alleles result in A and B antigens – chemicals on the surface of red blood cells.

The O type blood cells carry no antigen.

DISCRETE VARIATION | 49

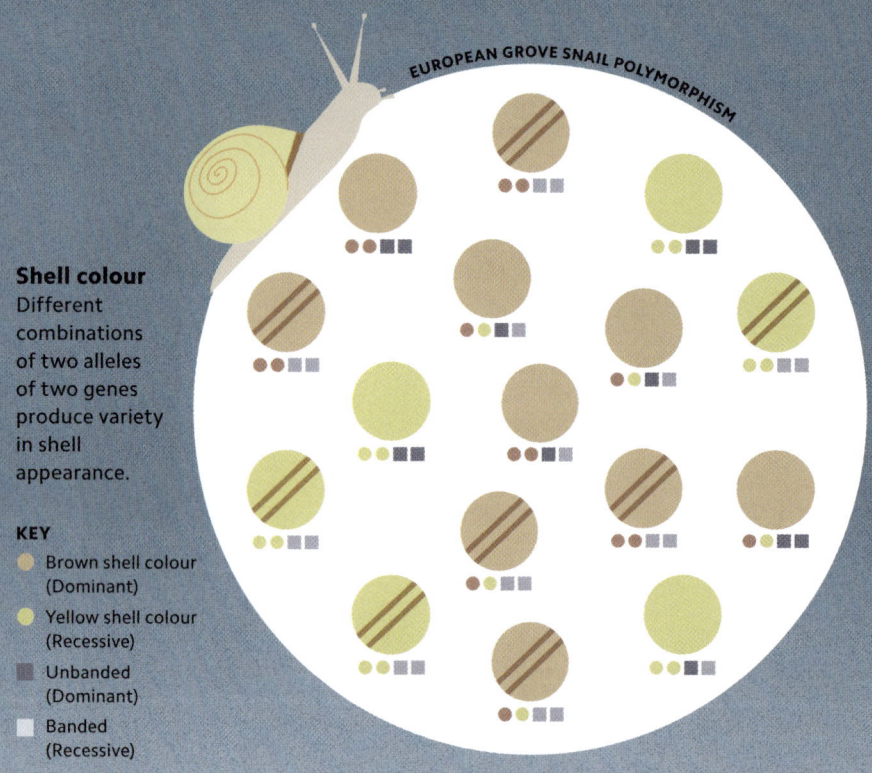

EUROPEAN GROVE SNAIL POLYMORPHISM

Shell colour
Different combinations of two alleles of two genes produce variety in shell appearance.

KEY
- Brown shell colour (Dominant)
- Yellow shell colour (Recessive)
- Unbanded (Dominant)
- Banded (Recessive)

GENES IN COMBINATION

Genes can interact with one another, altering their effects when in combination. This interaction can generate great variation in the characteristics of individuals in a population. Consider just two genes that influence one characteristic – the appearance of the shell of the European grove snail (*Cepaea nemoralis*). One gene determines if the shell is brown or yellow in colour, the other if it is banded or unbanded. Different combinations of the genes produce several phenotypes. In reality, shell colour is controlled by more than just two genes, producing many shell variants. Such genetic diversity in a population is called polymorphism.

A SLIDING SCALE

Some characteristics are controlled by very large numbers of genes. Human height is influenced by several hundred genes, some of which have a larger effect than others. Each new gene combination adds a new intermediate form until – effectively – the variants merge into a continuum, with no discrete height forms. Environmental factors, such as nutrition, also play a significant role in determining adult height.

High times
Human height varies continuously in individuals between extremes because it is the result of the activity of multiple genes and the environment.

BELL CURVE
A graph of human height frequency resembles a bell curve.

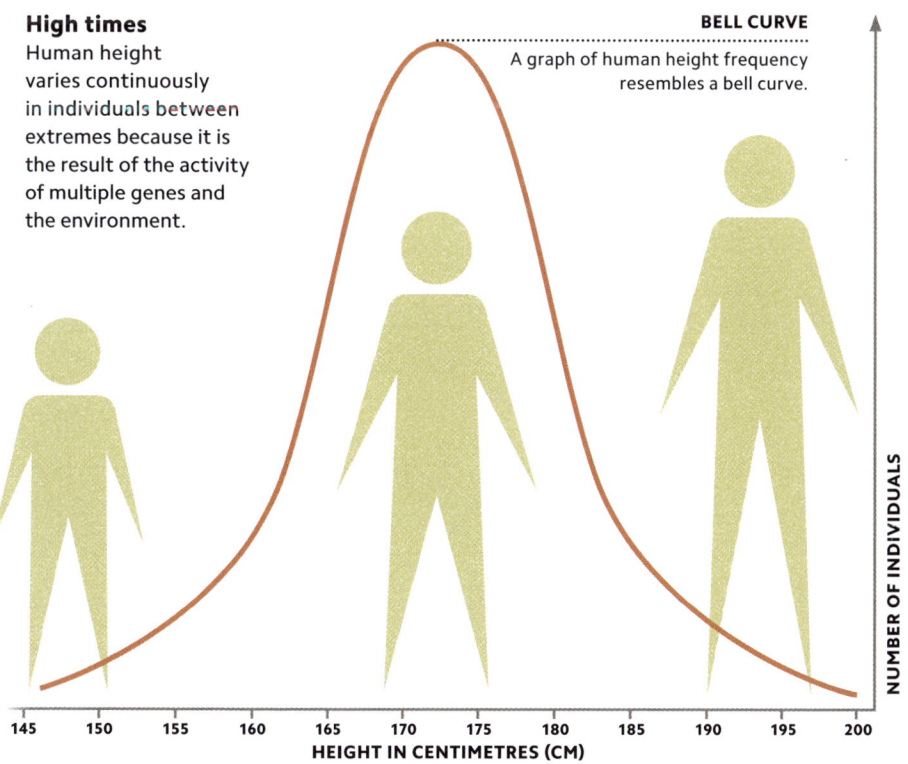

CONTINUOUS VARIATION | 51

GENE INTERACTION

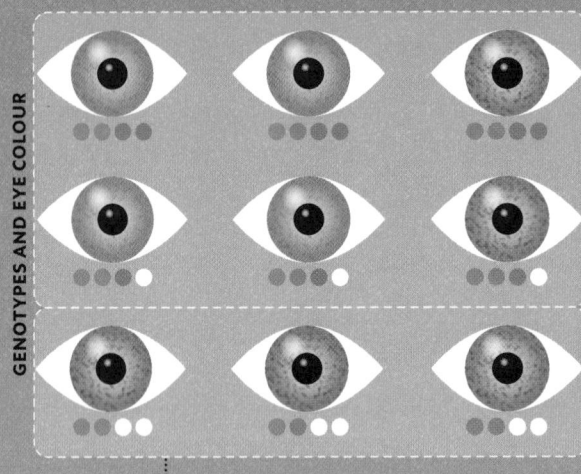

GENOTYPES AND EYE COLOUR

PIGMENT SWITCH GENE ON
The eye colour gene has two alleles; presence of the dominant allele produces brown pigment when the pigment gene is switched on.

PIGMENT SWITCH GENE OFF
The recessive form of the pigment switch gene stops pigment production, even when the dominant brown-eye allele is present.

TWO GENES CONTROLLING EYE COLOUR

PIGMENT GENE
- Brown pigment (dominant)
- No pigment (blue) (recessive)

PIGMENT SWITCH GENE
- Pigment on (dominant)
- Pigment off (recessive)

One gene can be moderated by the activity of another gene – a phenomenon called epistasis. For example, human eye colour is determined by two genes. One has two alleles that determine if the eye has brown pigment or has no pigment (leaving it blue). The other gene effectively switches the first gene on. If the pigment switch gene is off, the eyes will be blue.

EPISTASIS

Geranium
The form and colour of geranium leaves can be affected by both genetic and environmental factors. Too much water or too little light, for example, can turn green leaves yellow.

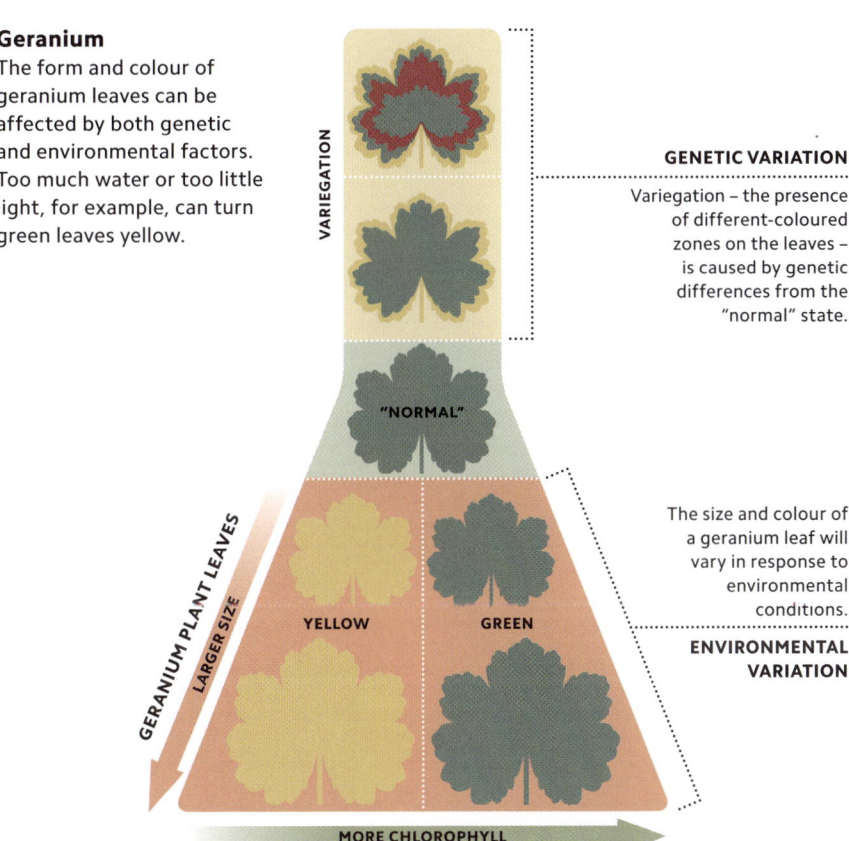

GENETIC VARIATION
Variegation – the presence of different-coloured zones on the leaves – is caused by genetic differences from the "normal" state.

The size and colour of a geranium leaf will vary in response to environmental conditions.

ENVIRONMENTAL VARIATION

OUTSIDE INFLUENCES

Not all the variation seen in living things is determined by genes. Many characteristics are moulded by their surroundings: green leaves turn pale when grown in shade, and skin tone darkens when exposed to the sun. This is because environmental factors such as nutrients, light, and temperature affect growth and development. Genes and the environment can also interact: in Siamese cats, for example, pigment-producing genes work only in the cooler extremities of the body, resulting in the cats having darker paws and ears.

ENVIRONMENTAL VARIATION

THE
DNA
MOLECU

The ability of living things to produce new life derives from the properties of DNA, a long chain molecule that holds a cell's genetic code. DNA is copied before a cell divides so that a complete set of genetic information is passed on to its daughter cells. DNA's famous double helix structure, which resembles a twisted ladder (see p.59), is key to the molecule's most important feature – it can act as a template for its own duplication. This allows multi-celled bodies to grow from a single fertilized egg and produce offspring of their own.

LADDERS OF LIFE

The DNA molecule is made up of just five elements – carbon (C), hydrogen (H), oxygen (O), nitrogen (N), and phosphorus (P). It belongs to a class of chemicals known as a polymers, which are long molecules made up of simpler, repeated subunits. In the case of DNA, these subunits are called nucleotides. Genetic material is slightly acidic and is found in high concentration within the cell nucleus; for these reasons DNA is termed a nucleic acid.

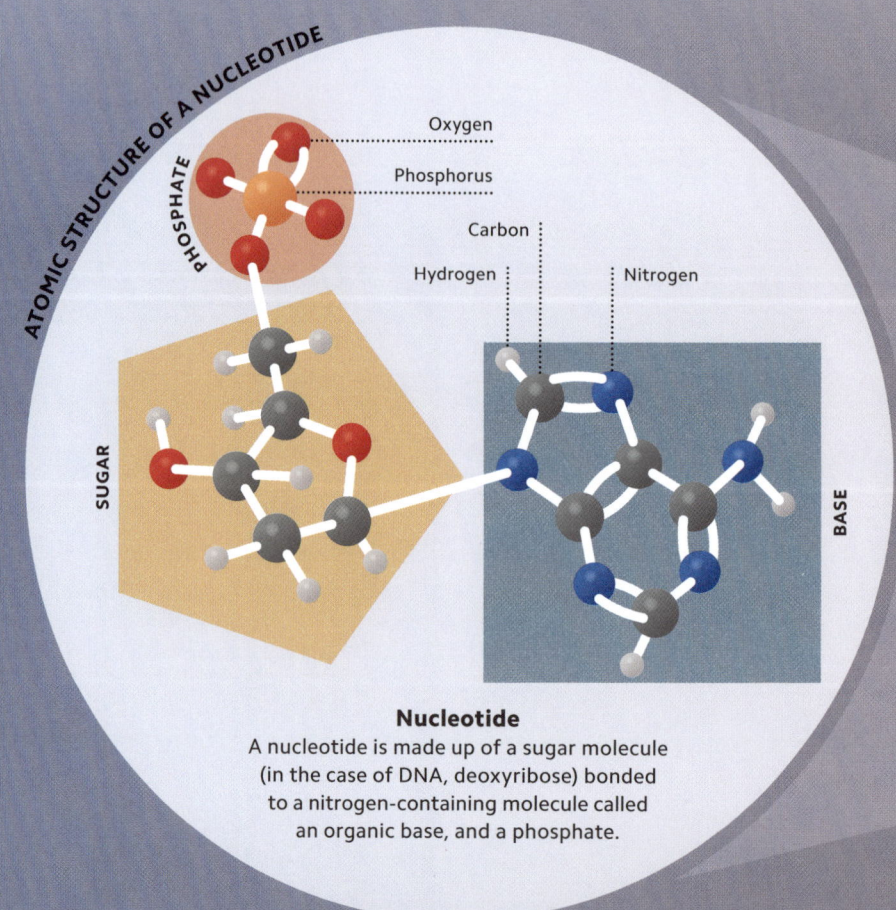

Nucleotide
A nucleotide is made up of a sugar molecule (in the case of DNA, deoxyribose) bonded to a nitrogen-containing molecule called an organic base, and a phosphate.

Chain molecule

The sugar and phosphate parts of nucleotides link up to form a long chain. The bases protrude from the chain and form weak bonds with the bases of a second, parallel, chain of nucleotides. This gives the DNA molecule the appearance of a ladder where the bases make up the "rungs".

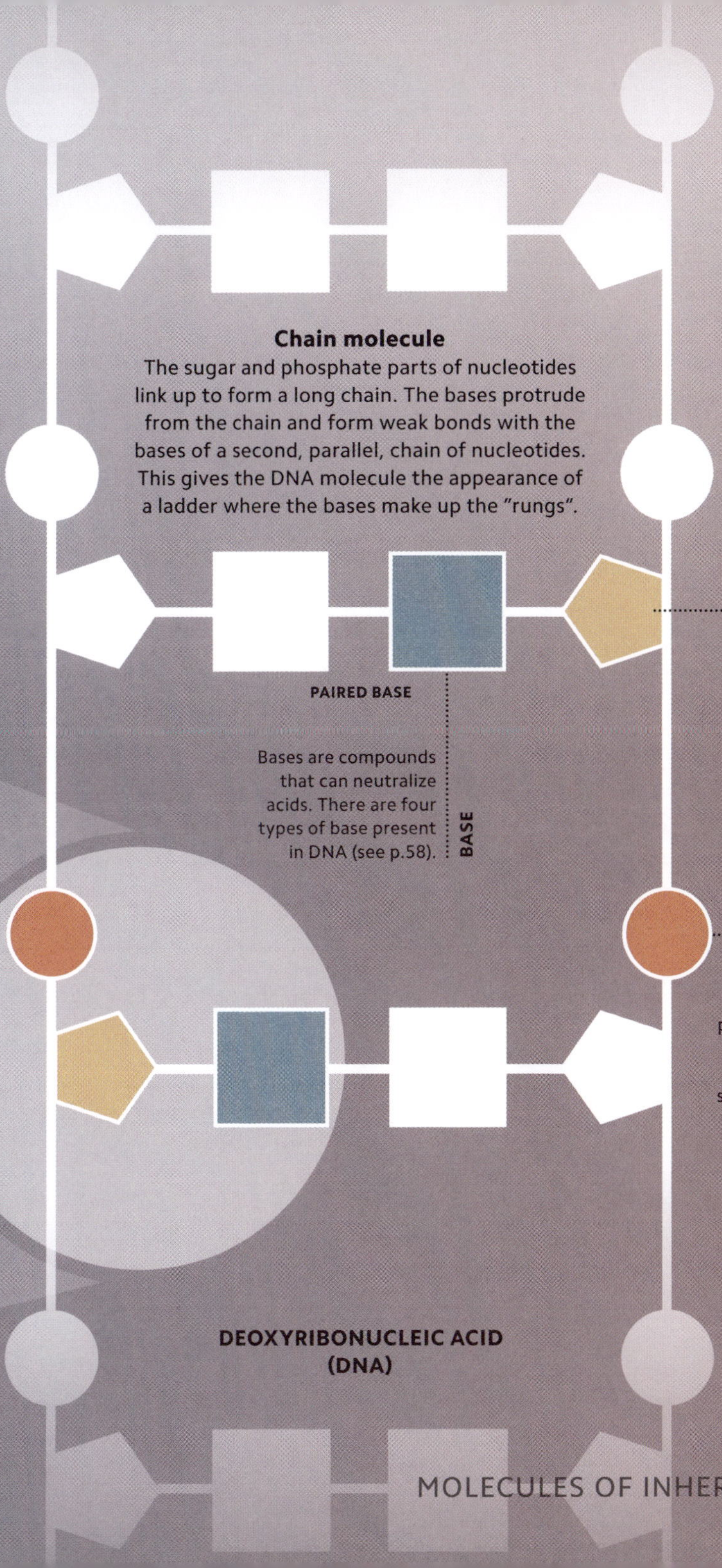

SUGAR

Sugars are simple carbohydrates. Deoxyribose, the sugar in DNA, contains five carbon atoms.

PAIRED BASE

BASE

Bases are compounds that can neutralize acids. There are four types of base present in DNA (see p.58).

PHOSPHATE

Phosphate groups are composed of phosphorus and oxygen. They alternate with sugar molecules to form the "backbone" of a DNA strand.

DEOXYRIBONUCLEIC ACID (DNA)

MOLECULES OF INHERITANCE

A TWIST OF FATE

The DNA molecule has a ladderlike structure, in which organic bases are the "rungs". There are four different bases – adenine (A), thymine (T), cytosine (C), and guanine (G). Crucially, they pair up in a specific way: adenine on one chain always pairs with thymine on the other, while guanine always bonds with cytosine. The configuration of these base pairs causes the ladder to become twisted, and take the form of two helixes twisted around one another – the double helix.

Base pairs
The base pairs A–T and C–G are held together by relatively weak chemical bonds called hydrogen bonds, which means that sections of DNA can be "unzipped" easily when needed.

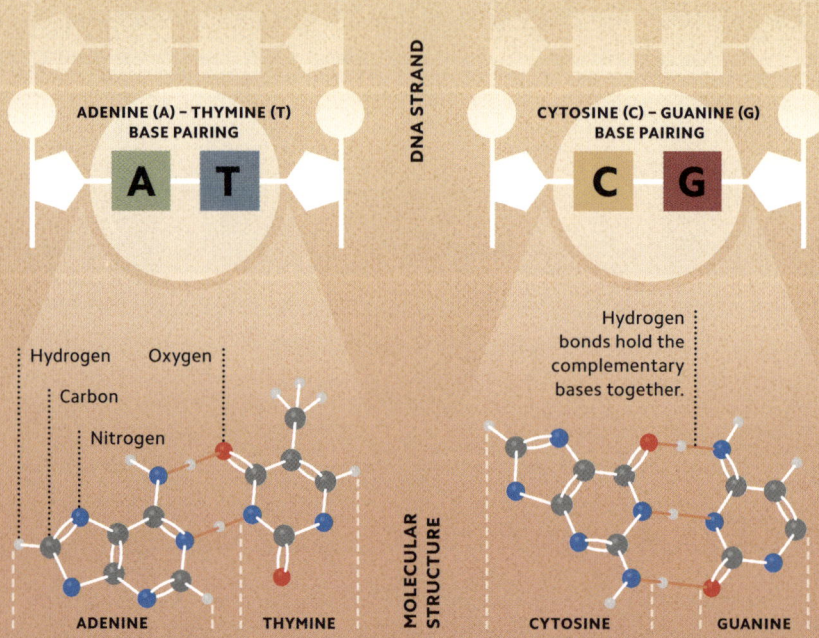

The double helix
The consistent A–T and C–G pairing means that one strand of the DNA molecule mirrors the other. This helps maintain the integrity of the genetic code.

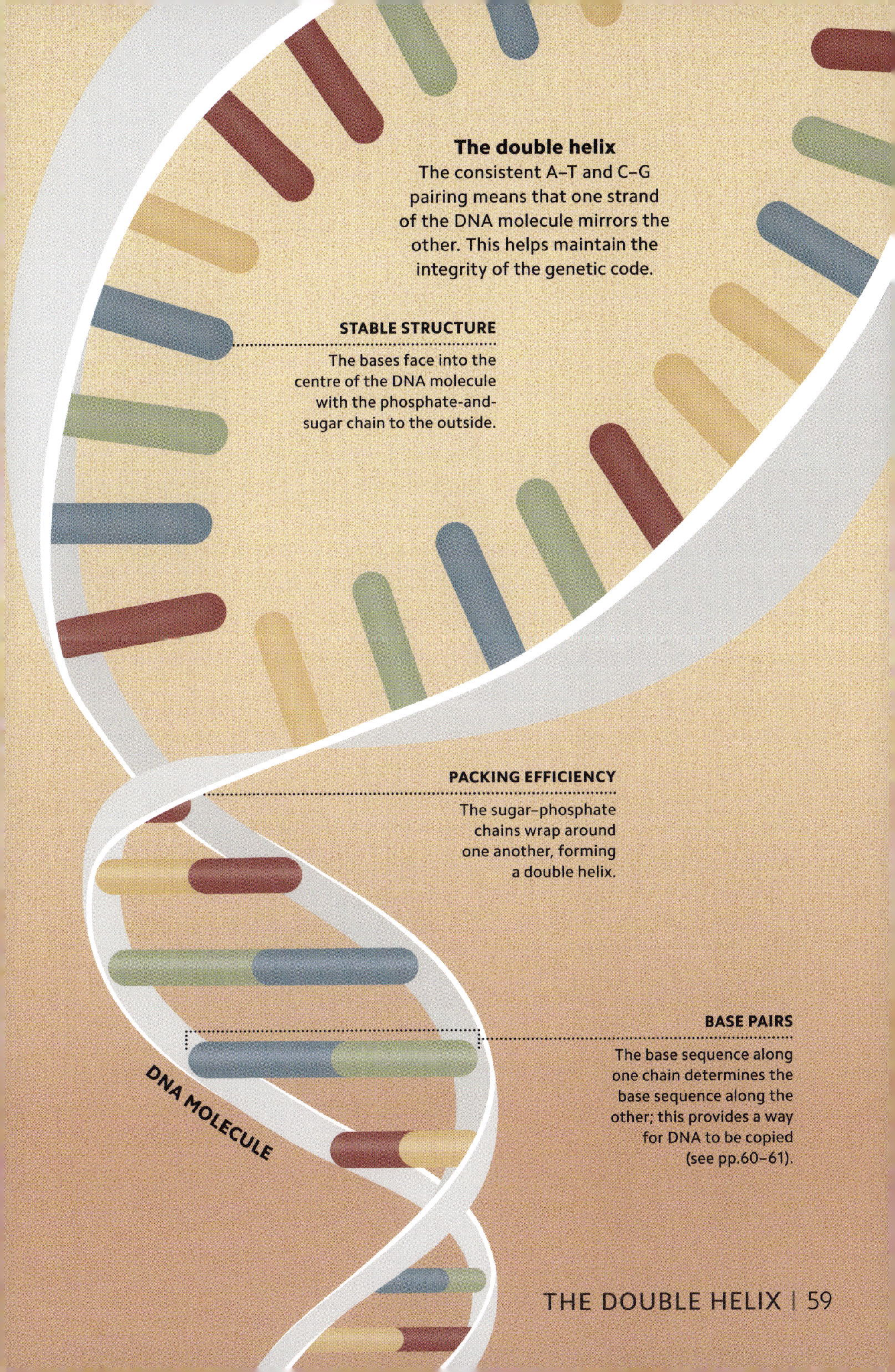

STABLE STRUCTURE
The bases face into the centre of the DNA molecule with the phosphate-and-sugar chain to the outside.

PACKING EFFICIENCY
The sugar–phosphate chains wrap around one another, forming a double helix.

BASE PAIRS
The base sequence along one chain determines the base sequence along the other; this provides a way for DNA to be copied (see pp.60–61).

DNA MOLECULE

THE DOUBLE HELIX | 59

MEMORY MOLECULES

Before a cell divides (see pp.68–69), its genetic material – DNA – is replicated. The structure of the molecule allows identical copies to be made, meaning that the code carried by DNA is passed reliably from one generation to the next. DNA is well suited to this job because it is made up of two strands, one of which is complementary to the other – adenine is always opposite thymine and guanine always opposite cytosine (see p.58). One strand of the DNA can therefore serve as a template against which the other strand can be assembled.

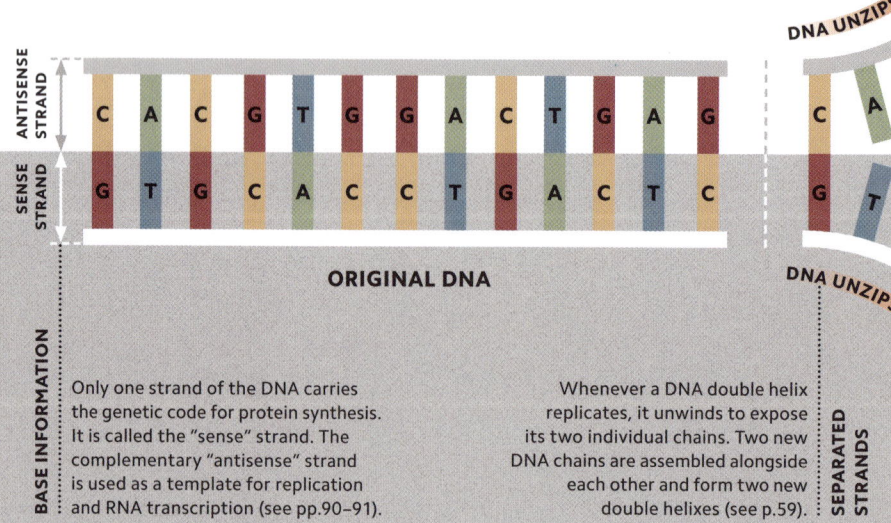

ANTISENSE STRAND / **SENSE STRAND**

C A C G T G G A C T G A G | C A
G T G C A C C T G A C T C | G T

ORIGINAL DNA

DNA UNZIPS / **DNA UNZIPS** / **SEPARATED STRANDS**

BASE INFORMATION: Only one strand of the DNA carries the genetic code for protein synthesis. It is called the "sense" strand. The complementary "antisense" strand is used as a template for replication and RNA transcription (see pp.90–91).

Whenever a DNA double helix replicates, it unwinds to expose its two individual chains. Two new DNA chains are assembled alongside each other and form two new double helixes (see p.59).

> A human cell can replicate all of its DNA in about one hour.

TEMPLATES AND REPLICATION

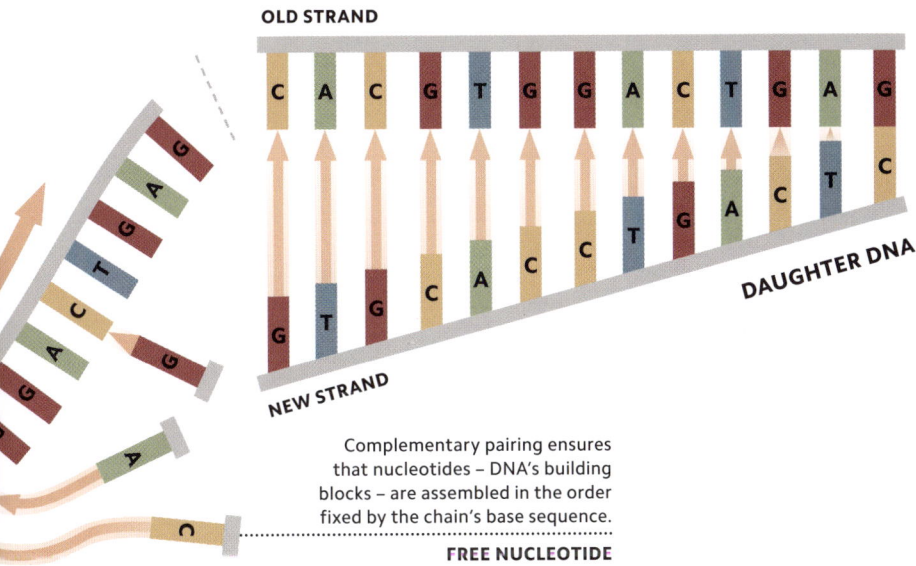

Complementary pairing ensures that nucleotides – DNA's building blocks – are assembled in the order fixed by the chain's base sequence.

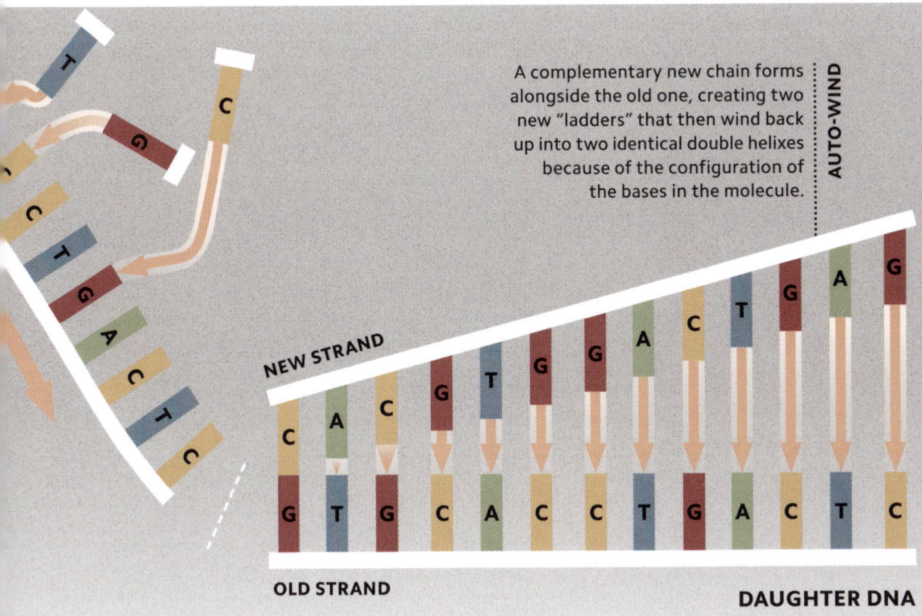

A complementary new chain forms alongside the old one, creating two new "ladders" that then wind back up into two identical double helixes because of the configuration of the bases in the molecule.

TEMPLATES AND REPLICATION

PACKED IN

The nucleus of a human cell holds 46 chromosomes, each of which contains one molecule of DNA. In some chromosomes, these molecules can be up to 4cm (1½in) long – 10,000 times longer than the cell itself. The DNA is well packaged; each double helix is coiled upon itself. The structure is held together with proteins called histones, forming a complex called chromatin.

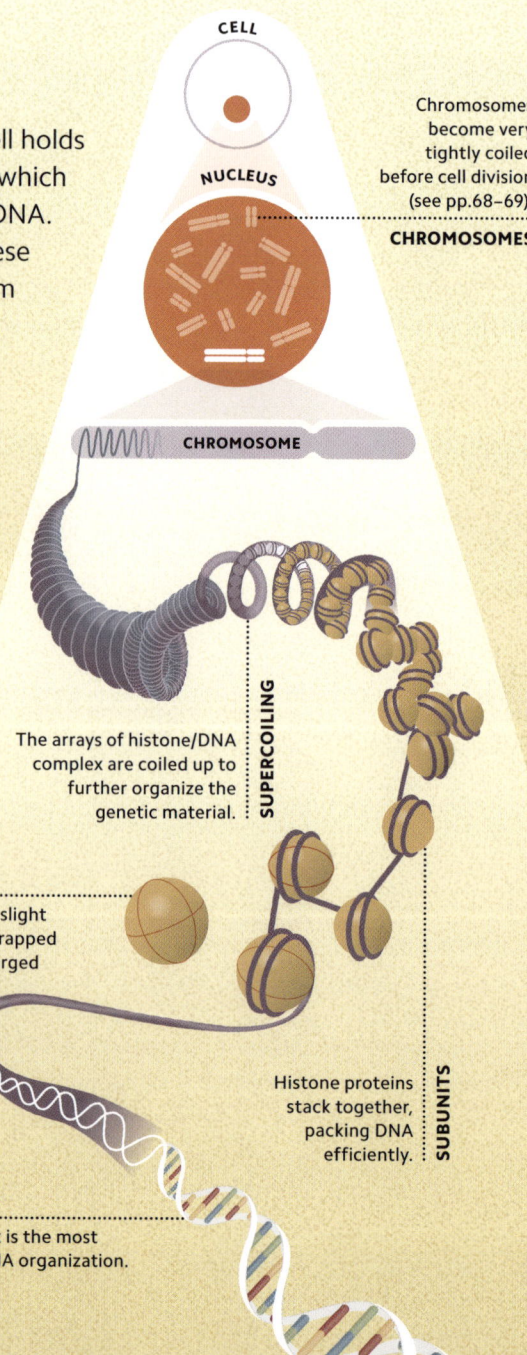

CELL

NUCLEUS

CHROMOSOMES

Chromosomes become very tightly coiled before cell division (see pp.68–69).

CHROMOSOME

SUPERCOILING

The arrays of histone/DNA complex are coiled up to further organize the genetic material.

HISTONE PROTEIN

DNA, which carries a slight negative charge, is wrapped around positively charged histone proteins.

SUBUNITS

Histone proteins stack together, packing DNA efficiently.

DOUBLE HELIX

The double helix is the most basic level of DNA organization.

62 | DNA COILING AND PACKAGING

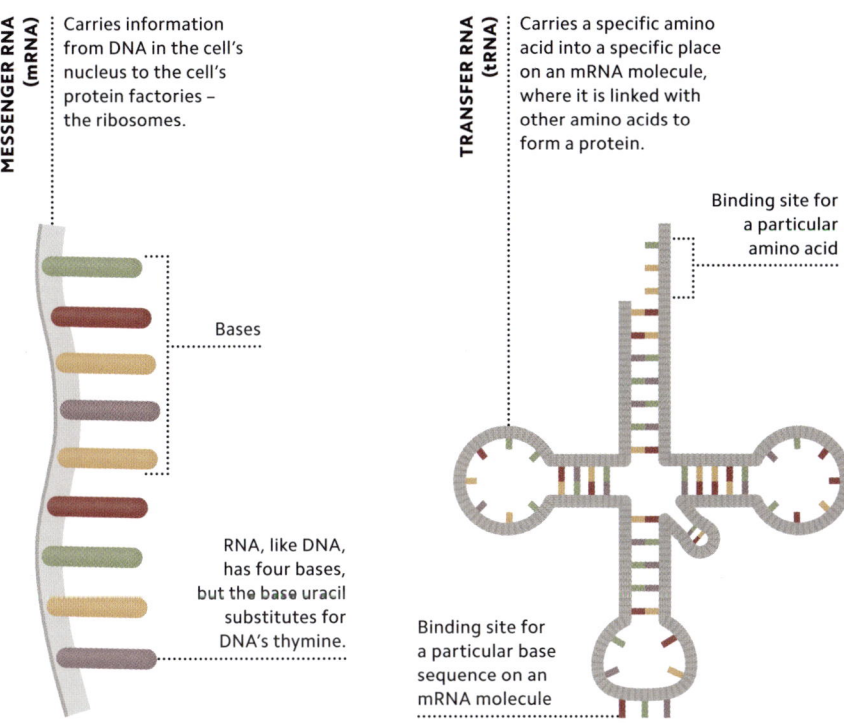

CELLULAR COURIERS

DNA is the genetic material in the nuclei of plant and animal cells. However, cells also contain another type of nucleic acid. Called ribonucleic acid (RNA), this molecule plays vital roles in executing the instructions encoded in DNA's structure. RNA molecules are shorter than those of DNA and are made up of a single (rather than double) chain of sugar (ribose rather than deoxyribose) and phosphate. There are two main types of RNA: messenger RNA (mRNA) and transfer RNA (tRNA). RNA also contributes to the structure of ribosomes – the sites in the cell where proteins are assembled (see pp.92–93).

THE 98 PER CENT

Perhaps surprisingly, only a small fraction of the DNA in a eukaryotic cell codes for proteins (see p.17). The rest – known as non-coding DNA (ncDNA) – is not "junk" as was once thought. It plays important roles in switching genes on and off (see pp.96–99); controlling gene function; and contributing to the structural parts of chromosomes, such as the centromeres (see p.68). DNA sequences called transposons have been identified that can "jump" between locations in the genome, sometimes resulting in gene mutations (see pp.102–111).

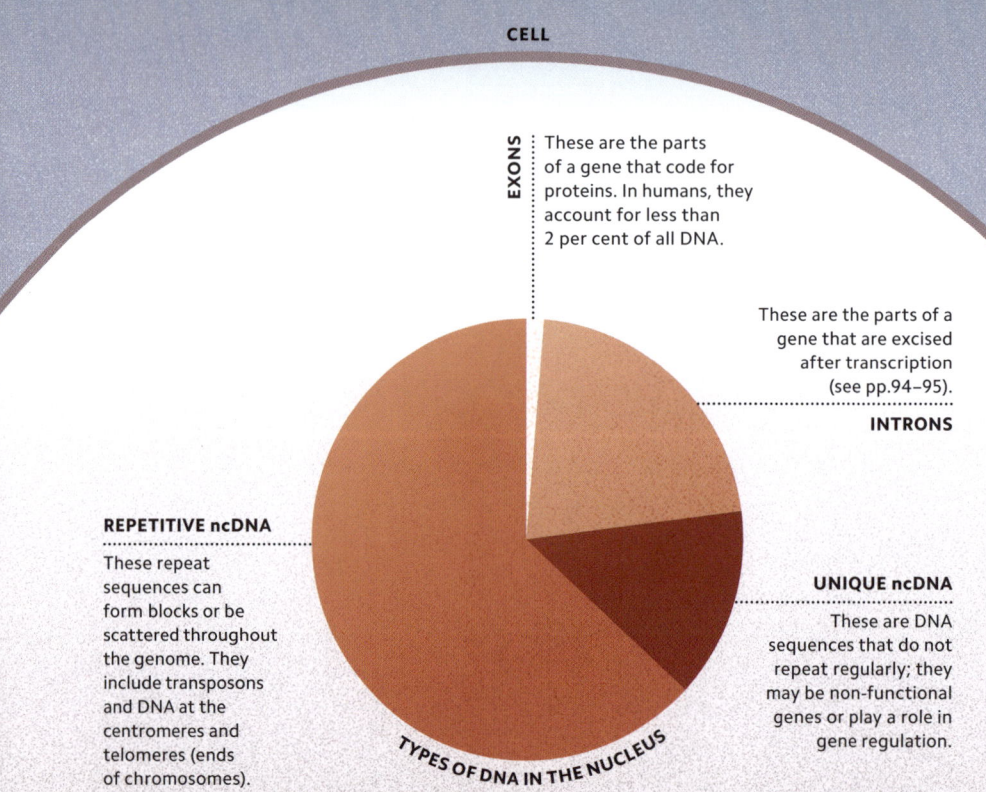

CELL

EXONS
These are the parts of a gene that code for proteins. In humans, they account for less than 2 per cent of all DNA.

INTRONS
These are the parts of a gene that are excised after transcription (see pp.94–95).

REPETITIVE ncDNA
These repeat sequences can form blocks or be scattered throughout the genome. They include transposons and DNA at the centromeres and telomeres (ends of chromosomes).

UNIQUE ncDNA
These are DNA sequences that do not repeat regularly; they may be non-functional genes or play a role in gene regulation.

TYPES OF DNA IN THE NUCLEUS

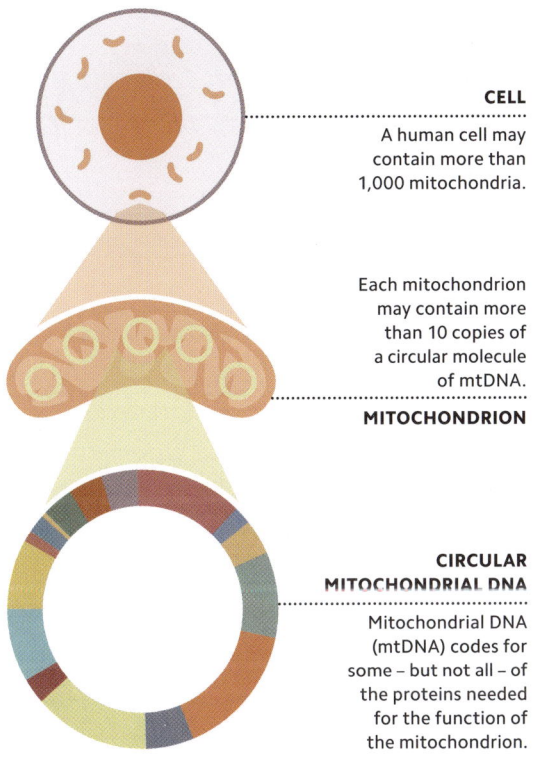

CELL
A human cell may contain more than 1,000 mitochondria.

Each mitochondrion may contain more than 10 copies of a circular molecule of mtDNA.

MITOCHONDRION

CIRCULAR MITOCHONDRIAL DNA

Mitochondrial DNA (mtDNA) codes for some – but not all – of the proteins needed for the function of the mitochondrion.

NON-NUCLEAR DNA

Most of the DNA in a eukaryotic cell (see p.15) resides in the chromosomes within the cell nucleus. Some, however, is found within the mitochondria – structures that produce the energy used to power the cell's biochemical processes. Mitochondria have a small genome of 37 genes on a circular molecule of DNA. This is a legacy from the ancestors of mitochondria – bacteria that settled inside our single-celled predecessors some 1.5 billion years ago. Mitochondria and their DNA (mtDNA) are inherited from the maternal parent.

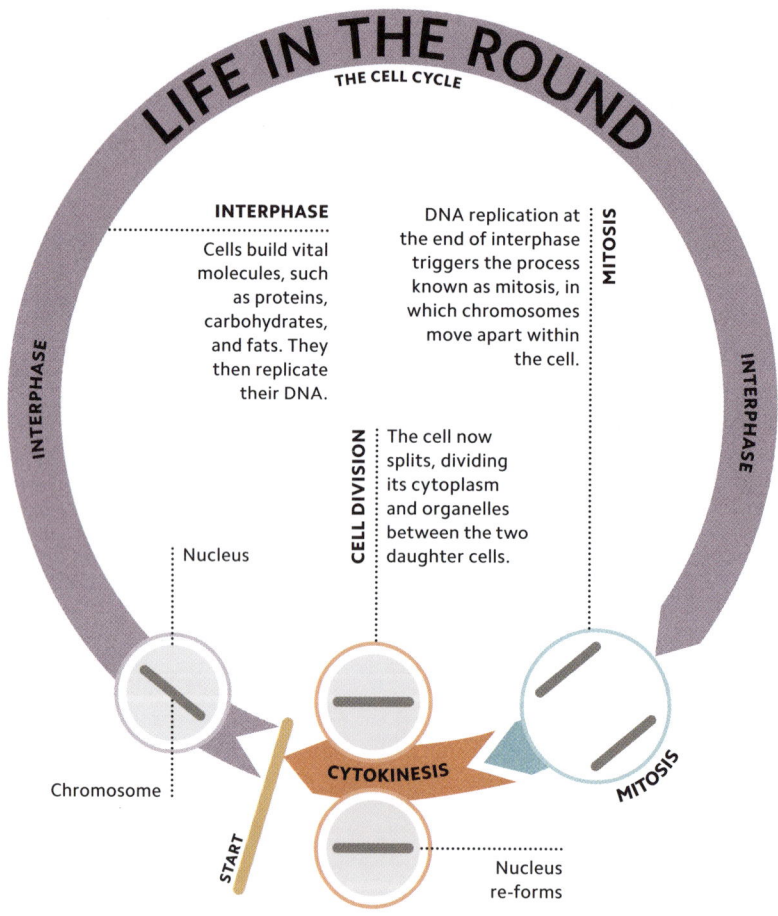

LIFE IN THE ROUND
THE CELL CYCLE

A cell's DNA is replicated just before a cell divides (see pp.68–69), ensuring that the two daughter cells contain all the genetic information they need to function. DNA replication and cell division are precisely controlled in what is called the cell cycle. The whole cycle typically lasts around 24 hours in human cells. Cells first go through a period of enlargement, called interphase: this is the longest part of the cycle, lasting many hours. This is followed by mitosis and then cell division, or cytokinesis, after which the cycle begins again.

DOUBLING DOWN

There are two types of cell division – mitosis and meiosis. Mitosis produces cells that are genetically identical to their parent cell. It provides the new cells required for the growth and repair of an organism. Mitosis happens in almost all cells of a growing body. A second type of cell division called meiosis occurs only in the sex organs; its function is to produce the sex cells, or gametes, such as the sperm and eggs.

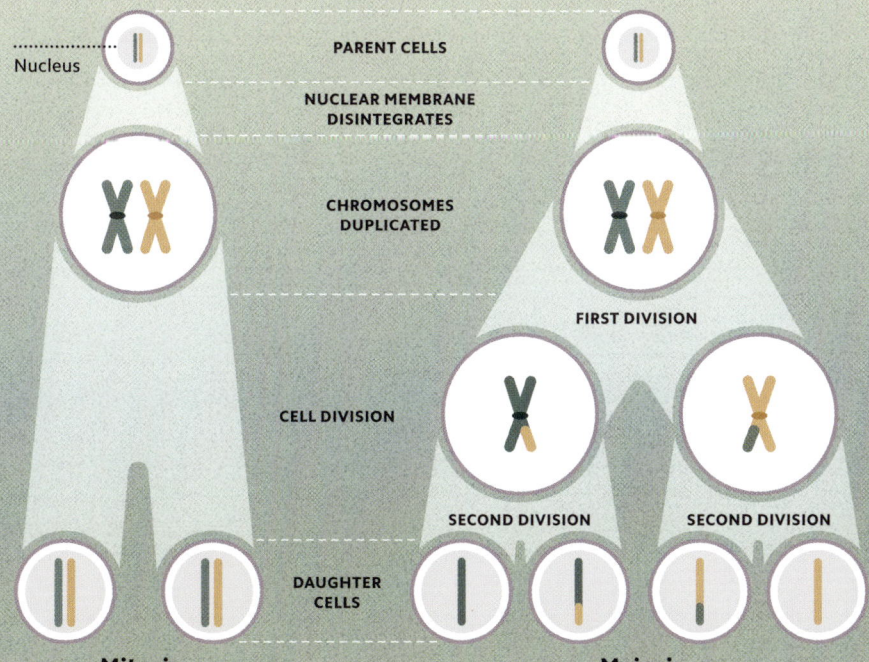

Mitosis
Two daughter cells are formed. They are genetically identical to the parent cell and have the same number of chromosomes.

Meiosis
Four daughter cells are formed. These are sex cells with half the number of chromosomes of the parent, and are genetically different from them (see pp.78–81).

Dance of the Chromosomes

Once a dividing cell has replicated its DNA (see pp.70–71), a precise sequence of cellular mechanics ensures that two daughter cells form and that each receives an identical full set of genetic material. This process, known as mitosis, can take less than an hour and may be observed under a microscope. It is orchestrated by structures within the cell, including long cablelike strands called microtubules that are able to shorten and lengthen rapidly.

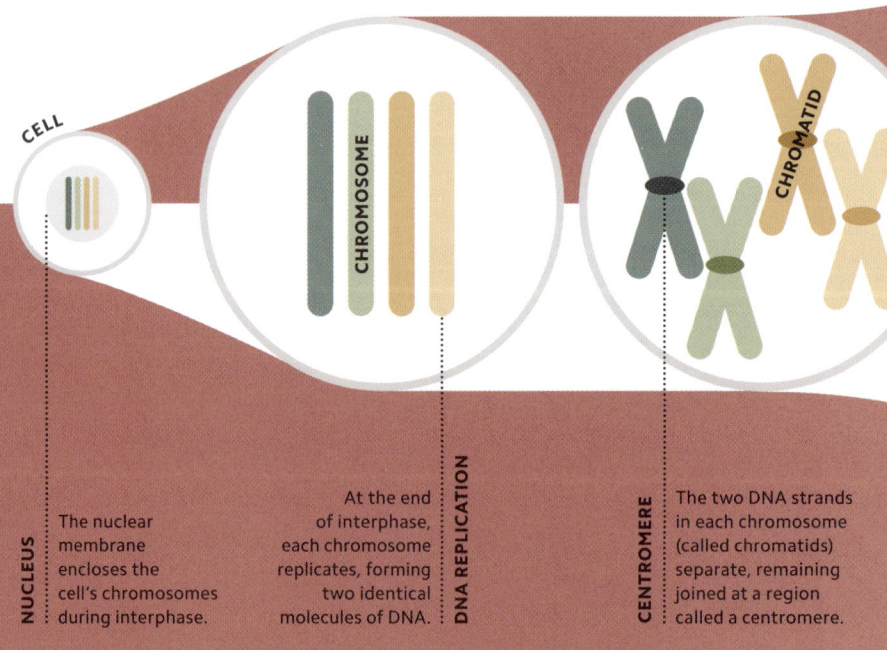

NUCLEUS — The nuclear membrane encloses the cell's chromosomes during interphase.

DNA REPLICATION — At the end of interphase, each chromosome replicates, forming two identical molecules of DNA.

CENTROMERE — The two DNA strands in each chromosome (called chromatids) separate, remaining joined at a region called a centromere.

Interphase
During interphase (see p.66), a cell grows, generates energy-rich compounds, and makes the molecules it will later need to become two separate cells.

Prophase
The cell's DNA condenses into thick chromatids, which can be seen under a microscope. The nuclear membrane dissolves.

PHASES OF MITOSIS

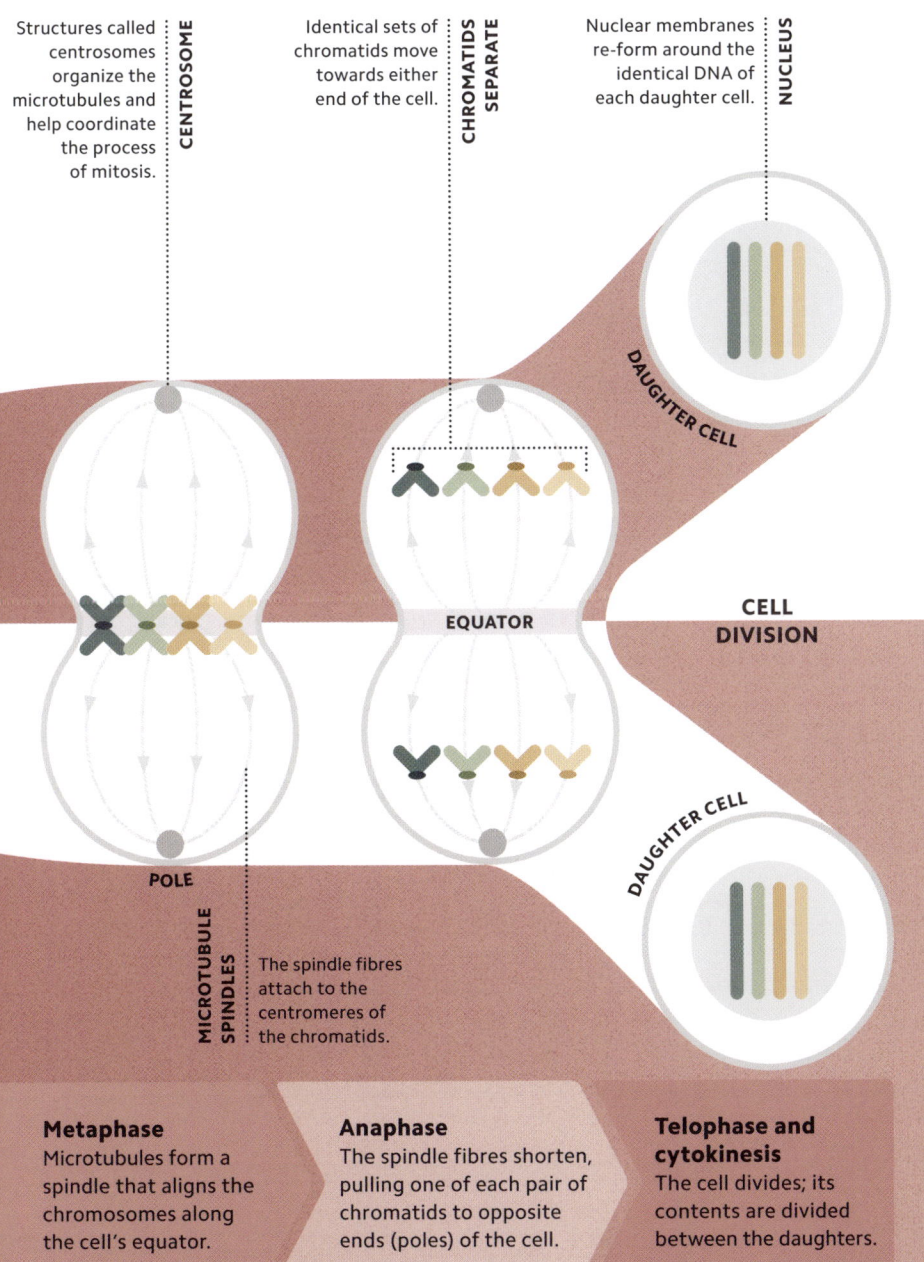

PHASES OF MITOSIS | 69

QUICK COPIES

A single human chromosome contains a molecule of DNA carrying tens of millions of base pairs – a vast amount of information to be copied every time a cell divides. Copying a DNA strand in one pass – from one end to the other – would be far too slow. Instead, replication takes place at many points along each DNA molecule. There are multiple "bubbles" where the two DNA strands unzip and separate, exposing templates against which new nucleotides are added.

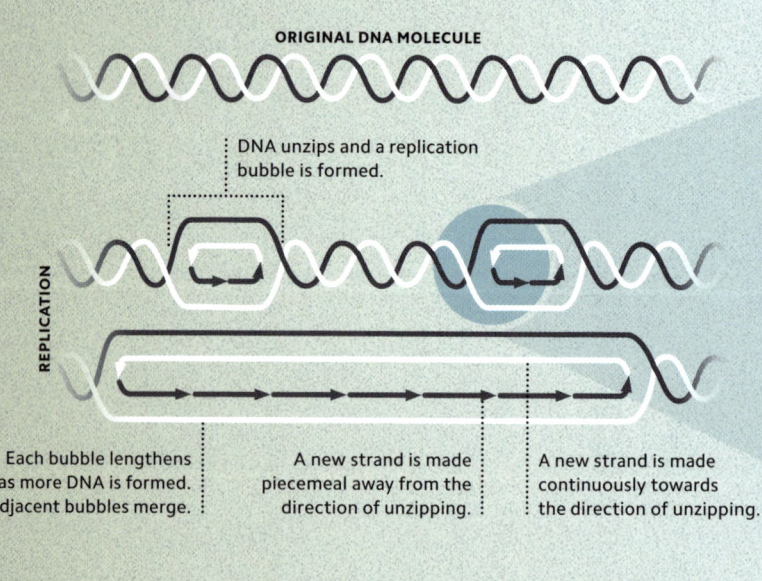

DNA unzips and a replication bubble is formed.

Each bubble lengthens as more DNA is formed. Adjacent bubbles merge.

A new strand is made piecemeal away from the direction of unzipping.

A new strand is made continuously towards the direction of unzipping.

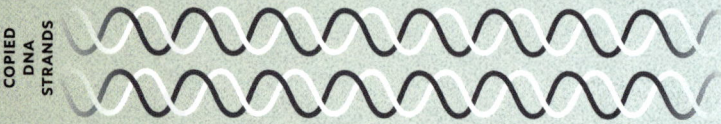

Replication bubbles
The many bubbles along a chromosome lengthen and eventually merge with their neighbours. Two identical DNA strands are the result.

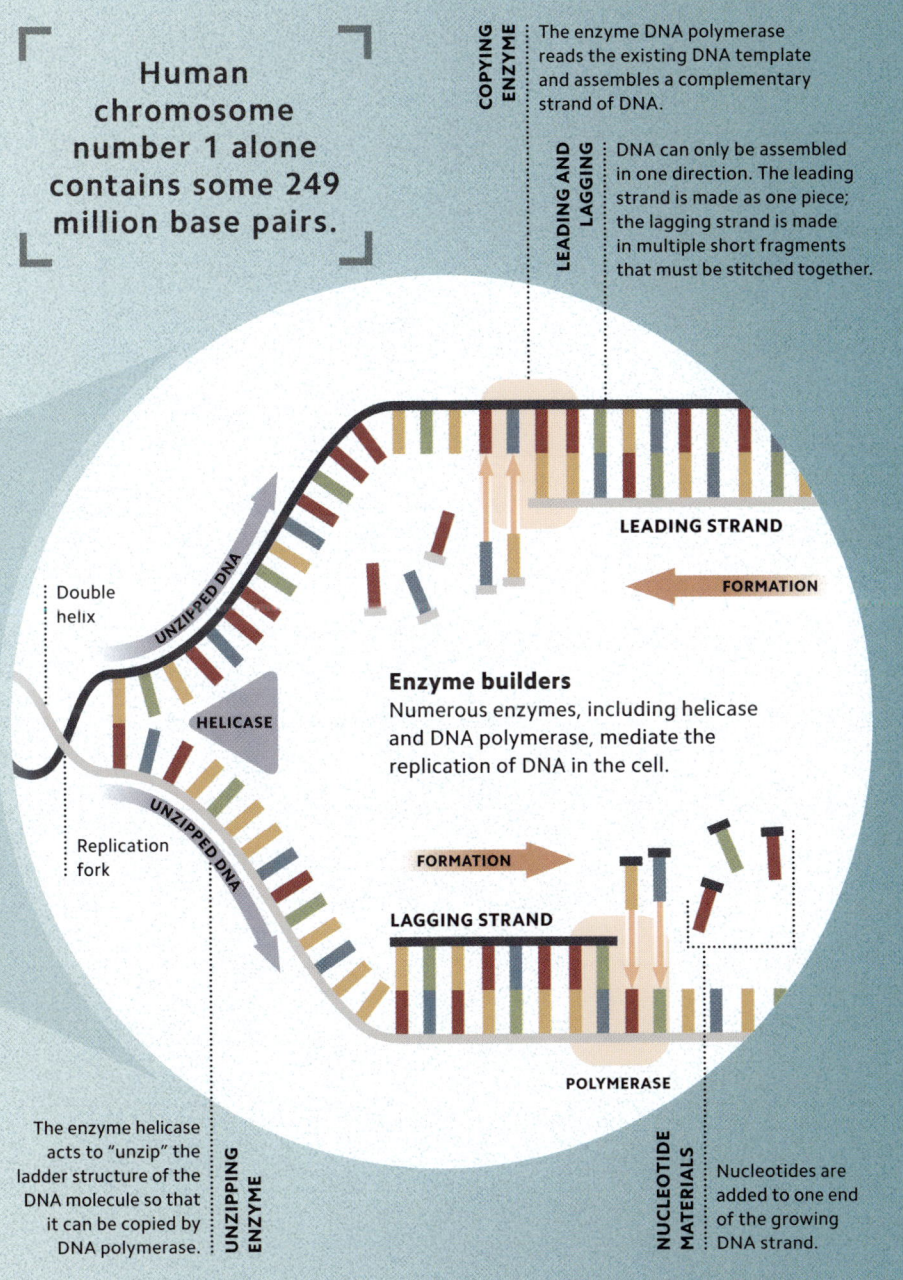

DNA REPLICATION

GENESA
REPROD

INTRODUCTION

Living things produce new generations of individuals to replace those that die. A few organisms can make genetically identical clones, but most have more complex life cycles involving mating between two parents. This mixes parental genes, helping to create genetic variety that in turn generates variety in physical and behavioural traits. This improves the chances of some offspring surviving in a changeable world. Genetic variety is produced by events that occur at a microscopic level – a special type of cell division called meiosis generates sex cells that fuse at fertilization and transmit genes from one generation to the next.

CHANGE FOR THE BETTER

The simplest way for an organism to reproduce is without sex. Bacteria and some plants reproduce in this manner, as do some animals; aphids, for example, create offspring from unfertilized eggs. However, sexual life cycles are favoured in most organisms even though mating involves cost and risk. This is because it produces variation in physical and behavioural characteristics that increases the chance of some offspring surviving in a changing world.

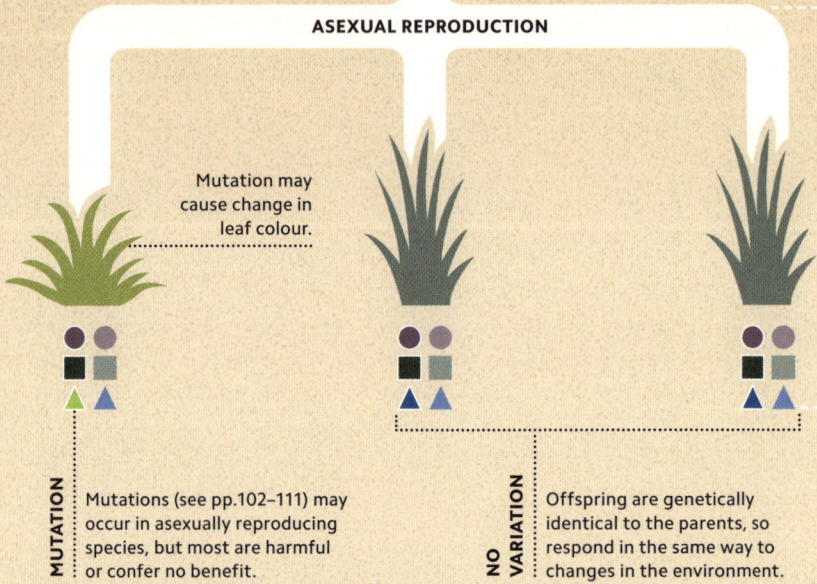

Parental genes are passed unchanged to the offspring.
PARENTAL GENOTYPE

Asexual reproduction
Many plants can reproduce quickly and prolifically without the need for sex cells. Some do this by producing shoots that root and detach, growing into independent offspring.

ASEXUAL REPRODUCTION

Mutation may cause change in leaf colour.

MUTATION — Mutations (see pp.102–111) may occur in asexually reproducing species, but most are harmful or confer no benefit.

NO VARIATION — Offspring are genetically identical to the parents, so respond in the same way to changes in the environment.

74 | GENERATING GENETIC DIVERSITY

KEY

GRASS GENES
● ■ ▲ Dominant
● ■ ▲ Recessive

RABBIT GENES
● ■ ▲ Dominant
● ■ ▲ Recessive

Sexual reproduction

This involves the production of male and female sex cells and their fusion in fertilization. These processes reshuffle the parental genes, resulting in genetically different offspring. Some of the offspring will be better adapted to changes in the environment than others and their genes are more likely to be passed on to subsequent generations.

PARENT(S)

SEXUAL REPRODUCTION

OFFSPRING

Mutation may cause change in fur colour.

Longer ears may help to radiate heat in a warming environment.

Shorter body length may make the rabbit slower to evade predators.

MUTATION
Variation can occasionally arise through mutation. Some mutations are beneficial to the individual, but most are harmful or neutral.

RESHUFFLING VARIATION
Some variations caused by the sexual reshuffling of genes improve the survival chances of the offspring; others do not.

GENERATING GENETIC DIVERSITY

STATES OF BEING

Most plants and animals have two sets of chromosomes, and therefore of genes, in their body cells. This double complement of chromosomes is described as diploid. They produce sex cells – eggs, sperm, or pollen nuclei – that contain a single set of chromosomes (and therefore genes) and are known as haploid. In some fungi and plants, however, the main life stage is haploid rather than diploid.

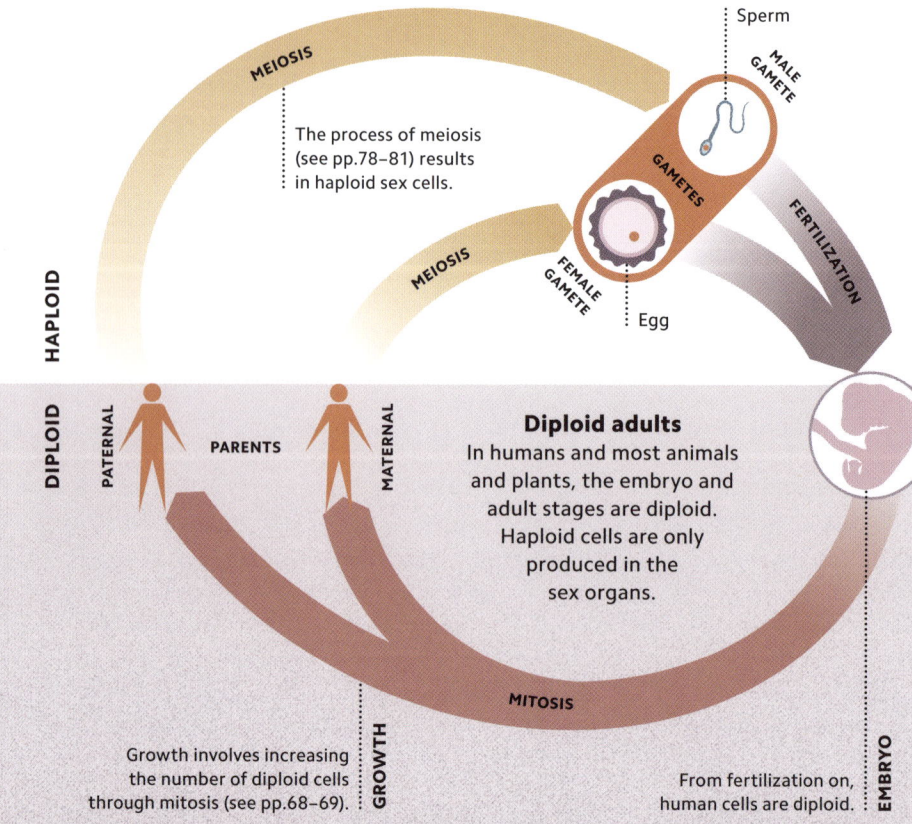

The process of meiosis (see pp.78–81) results in haploid sex cells.

Diploid adults
In humans and most animals and plants, the embryo and adult stages are diploid. Haploid cells are only produced in the sex organs.

Growth involves increasing the number of diploid cells through mitosis (see pp.68–69).

From fertilization on, human cells are diploid.

The moss life cycle alternates between diploid and haploid generations.

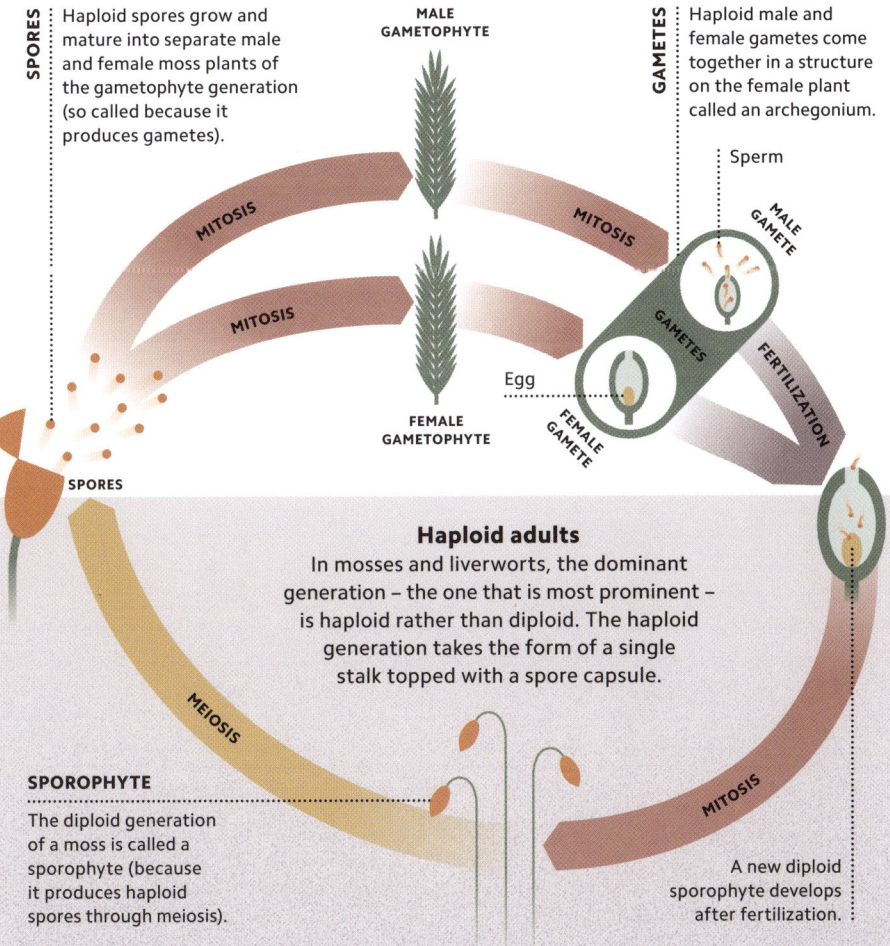

SPORES — Haploid spores grow and mature into separate male and female moss plants of the gametophyte generation (so called because it produces gametes).

GAMETES — Haploid male and female gametes come together in a structure on the female plant called an archegonium.

Haploid adults
In mosses and liverworts, the dominant generation – the one that is most prominent – is haploid rather than diploid. The haploid generation takes the form of a single stalk topped with a spore capsule.

SPOROPHYTE — The diploid generation of a moss is called a sporophyte (because it produces haploid spores through meiosis).

A new diploid sporophyte develops after fertilization.

DIPLOID AND HAPLOID | 77

REDUCTION DIVISION

The special kind of cell division that splits diploid cells into haploid eggs or sperm is called meiosis, from the Greek word for "lessening". Like mitosis (see pp.68–69), meiosis is preceded by DNA replication (see pp.70–71). What then follows are two rounds of cell division – meiosis I and II: the first halves the chromosome number, the second separates the DNA copies made during replication. Unlike mitosis, meiosis can reshuffle the alleles of the parents through a process called crossing over – where homologous chromosomes exchange genetic material.

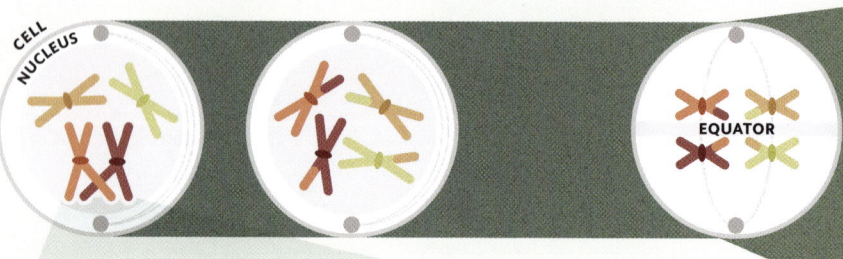

Crossing over
Sections of DNA can be swapped between homologues (see pp.42–43). Recombination occurs through crossing over.

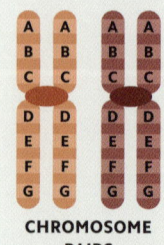
CHROMOSOME PAIRS

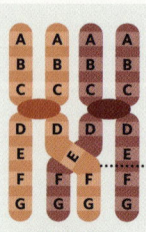
BRIDGED HOMOLOGUES

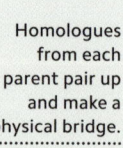
Homologues from each parent pair up and make a physical bridge.

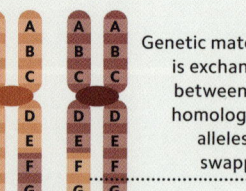
RECOMBINED CHROMOSOMES

Genetic material is exchanged between the homologues; alleles are swapped.

Prophase I
Chromosomes condense and the nucleus breaks down. Recombination between homologues occurs through crossing over.

Metaphase I
Homologues move to the equator of the cell and become attached to a spindle of microtubules (see pp.68–69).

CENTROSOME Structures called centrosomes organize the microtubules and help coordinate the process of meiosis.

> Meiosis takes place only in sex organs that produce gametes or spores.

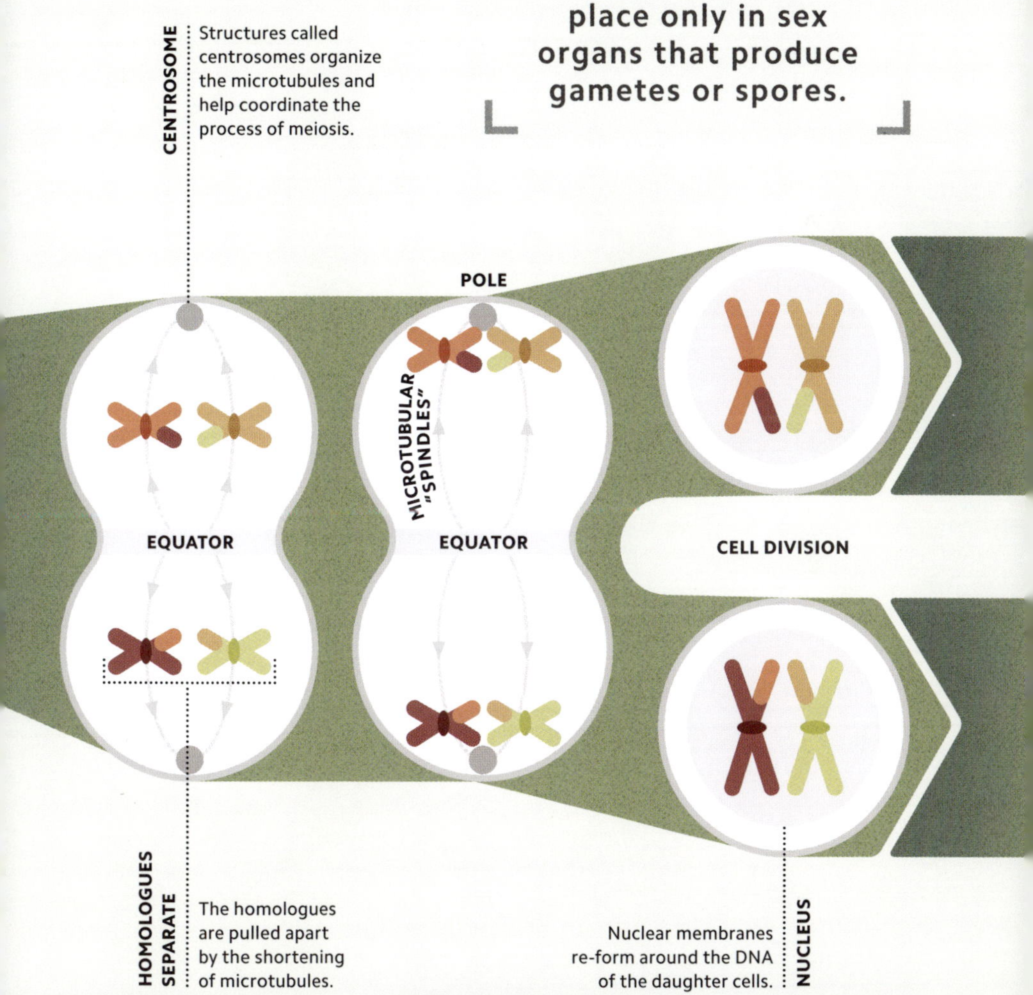

POLE

MICROTUBULAR "SPINDLES"

EQUATOR

EQUATOR

CELL DIVISION

HOMOLOGUES SEPARATE The homologues are pulled apart by the shortening of microtubules.

NUCLEUS Nuclear membranes re-form around the DNA of the daughter cells.

Anaphase I
Homologous chromosomes move to opposite ends (poles) of the cell.

Telophase I and cytokinesis
Chromosomes reach the poles of the cell; the cytoplasm divides, pinching the cell into two.

MEIOSIS I | 79

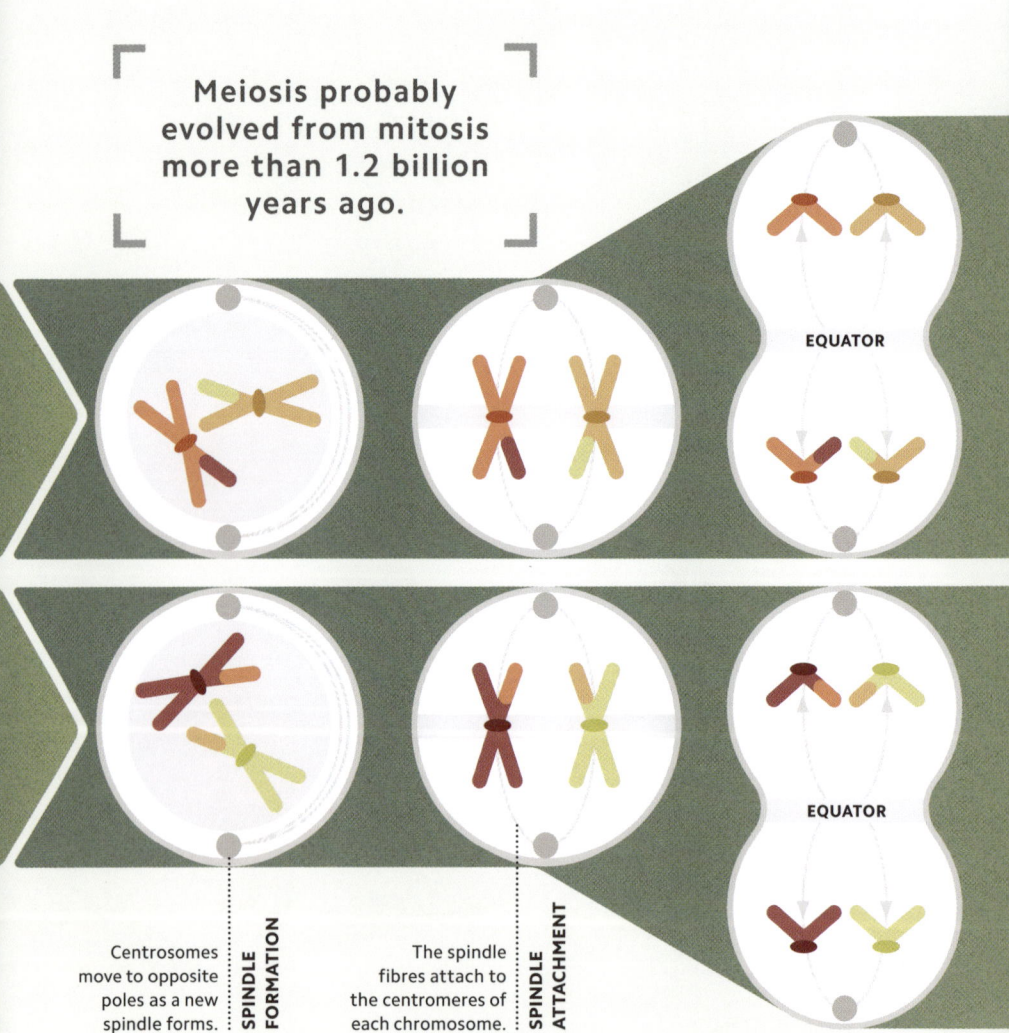

FORMING GAMETES

In the second part of meiosis, sister chromatids are separated and the cells formed by this division become the gametes, or sex cells. Eggs, made in female ovaries, are the larger sex cells and provide the bulk of cellular material for the first cell of the next generation. Sperm, made in male testes, are smaller and more plentiful. Having many male cells increases the chance of them meeting an egg. The haploid gene sets of male and female sex cells combine to restore the diploid set in the fertilized egg, or zygote.

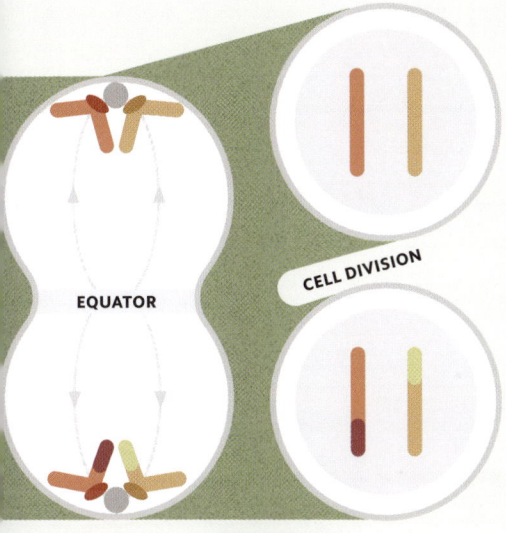

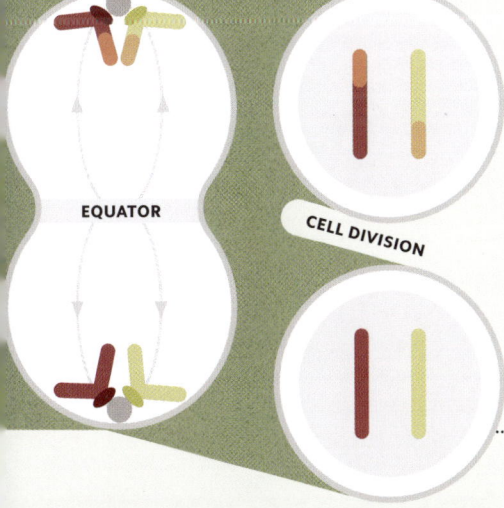

Telophase II and cytokinesis
Nuclear envelopes form, enclosing the chromosomes in each of the four separated haploid daughter cells.

HAPLOID CELLS
The sex cells formed by meiosis are haploid, and their genetic material has been reshuffled, producing offspring that differ from each other and both parents due to genetic recombination.

HOW GENES WORK

A living body is a complex machine. It combines the universal properties of life – such as growth, respiration, and reproduction – with specializations that define the individual and its species. All of these characteristics emerge from the working of genes that regulate the development and function of cells. This regulation involves complex interactions between biomolecules. Genes are sections of DNA that carry digital information; this information is used to make proteins that drive vital processes and build the body. Any one gene codes for one particular protein.

ONE GENE, ONE PROTEIN

Techniques in cell biology developed in the 20th century helped to unravel how genes worked at a molecular level. The first clues came from experiments on bread mould (*Neurospora*), which showed that a single gene mutation affected a single chemical pathway in the cell. Each pathway is catalysed by an enzyme (a type of protein), leading to the hypothesis that one gene codes for the production of one enzyme or protein.

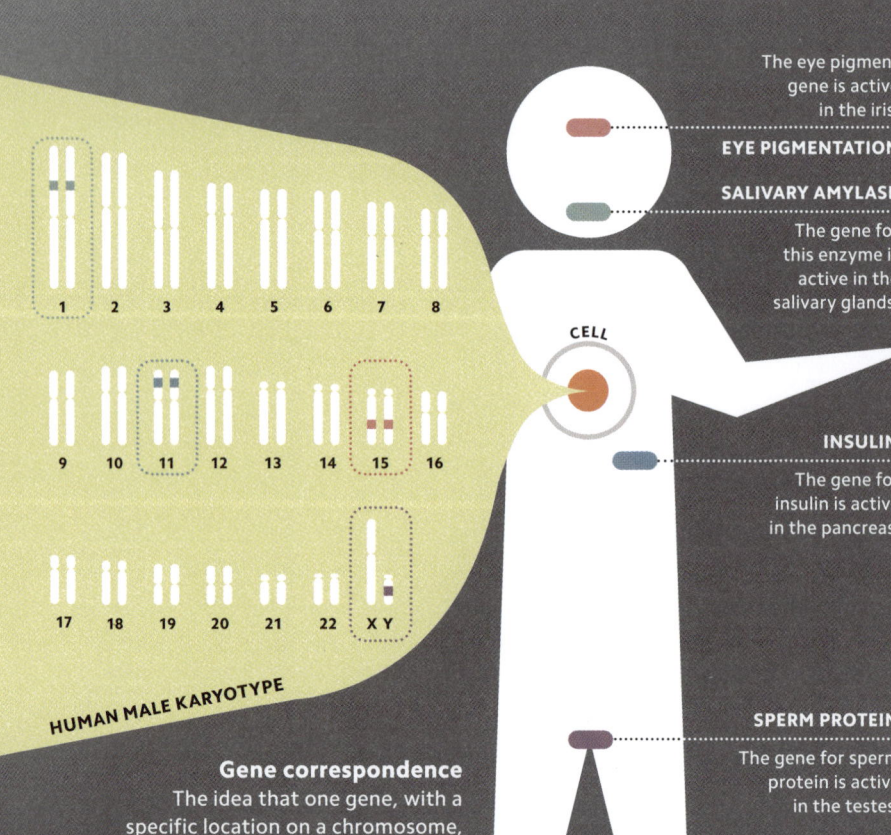

HUMAN MALE KARYOTYPE

EYE PIGMENTATION
The eye pigment gene is active in the iris.

SALIVARY AMYLASE
The gene for this enzyme is active in the salivary glands.

CELL

INSULIN
The gene for insulin is active in the pancreas.

SPERM PROTEIN
The gene for sperm protein is active in the testes.

Gene correspondence
The idea that one gene, with a specific location on a chromosome, codes for one protein is broadly correct.

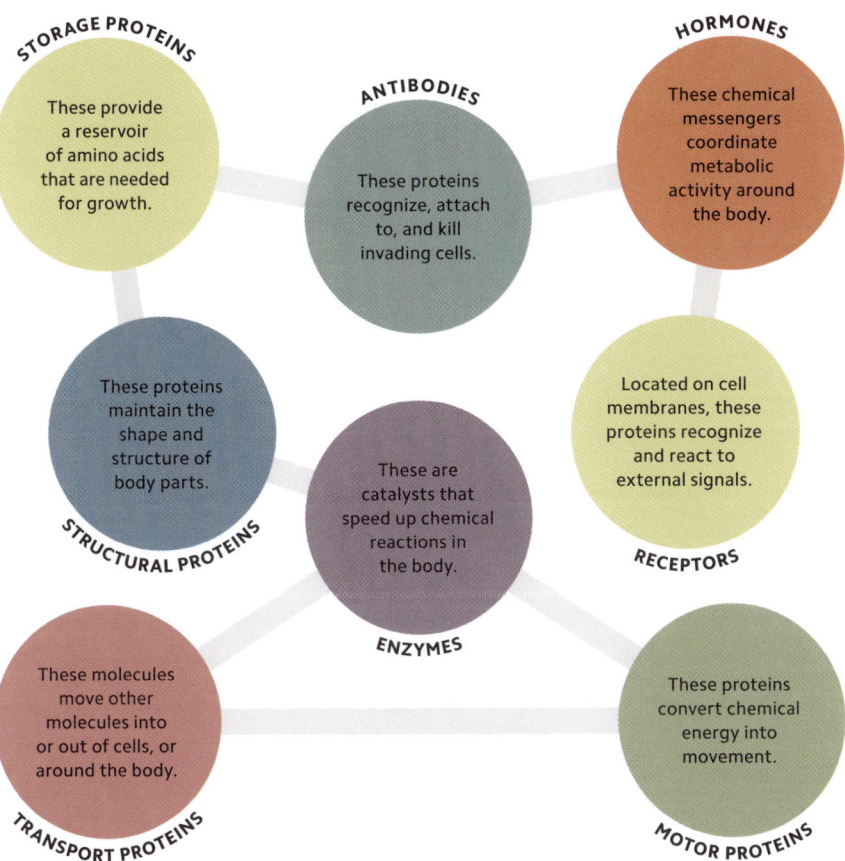

MOLECULES OF LIFE

Proteins are essential to life and have a huge variety of roles in the body. Some, such as collagen, are long-lived structural molecules, essential to skin, bone, and muscle. Others, such as insulin, break down after a few minutes, meaning that their levels in the body can be precisely controlled. Not all proteins are made by the body at the same time: genes can switch the production of proteins on and off.

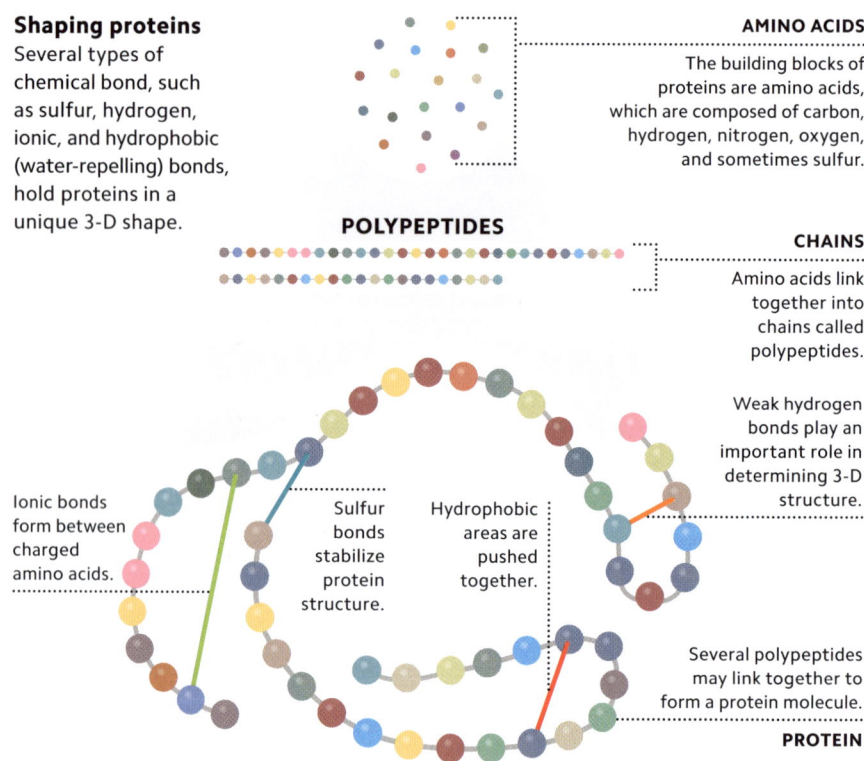

Shaping proteins
Several types of chemical bond, such as sulfur, hydrogen, ionic, and hydrophobic (water-repelling) bonds, hold proteins in a unique 3-D shape.

AMINO ACIDS
The building blocks of proteins are amino acids, which are composed of carbon, hydrogen, nitrogen, oxygen, and sometimes sulfur.

POLYPEPTIDES

CHAINS
Amino acids link together into chains called polypeptides.

Ionic bonds form between charged amino acids.

Sulfur bonds stabilize protein structure.

Hydrophobic areas are pushed together.

Weak hydrogen bonds play an important role in determining 3-D structure.

Several polypeptides may link together to form a protein molecule.

PROTEIN

FORM AND FUNCTION

Proteins are so diverse because their molecules are large and structurally complex. Every protein is made up of a long chain of amino acids – smaller organic molecules – linked together by so-called peptide bonds. Only 20 different amino acids occur in the proteins of living things, but they may be linked together into chains of many thousand units. The linear sequence of amino acids in a chain determines how the protein folds up on itself, giving it a unique three-dimensional shape. The function of the protein derives from its shape.

BIOLOGICAL CATALYSTS

Enzymes are a class of protein without which life could not exist. Their role is to speed up chemical reactions in the body that would otherwise take place extremely slowly. Cells contain many thousands of different enzymes; and the presence or absence of particular enzymes determines which chemical reactions take place within that cell. Enzymes show high specificity – they work only on one type of molecule – and although they control cellular reactions, they are not themselves consumed in those reactions.

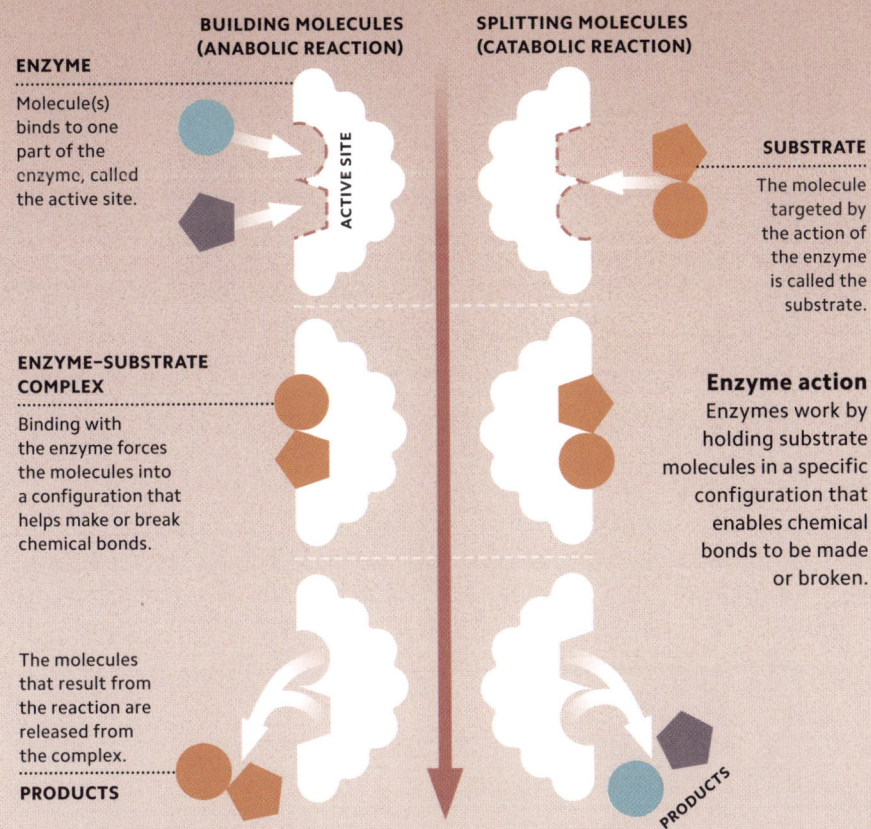

BUILDING MOLECULES (ANABOLIC REACTION)

SPLITTING MOLECULES (CATABOLIC REACTION)

ENZYME
Molecule(s) binds to one part of the enzyme, called the active site.

ACTIVE SITE

SUBSTRATE
The molecule targeted by the action of the enzyme is called the substrate.

ENZYME–SUBSTRATE COMPLEX
Binding with the enzyme forces the molecules into a configuration that helps make or break chemical bonds.

Enzyme action
Enzymes work by holding substrate molecules in a specific configuration that enables chemical bonds to be made or broken.

The molecules that result from the reaction are released from the complex.

PRODUCTS

PRODUCTS

PROTEINS AND ENZYMES

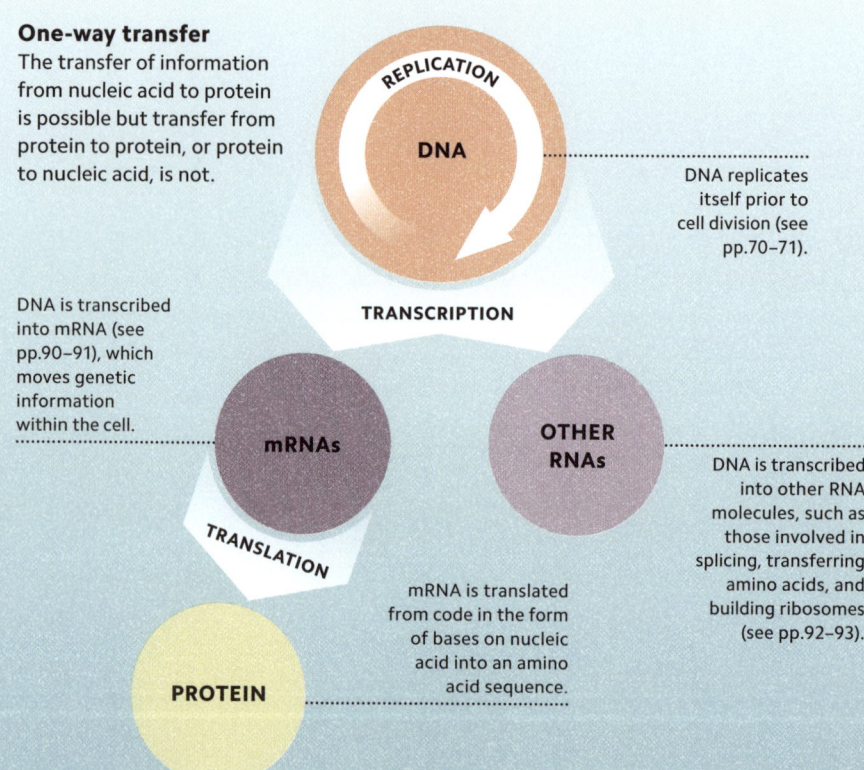

One-way transfer
The transfer of information from nucleic acid to protein is possible but transfer from protein to protein, or protein to nucleic acid, is not.

DNA replicates itself prior to cell division (see pp.70–71).

DNA is transcribed into mRNA (see pp.90–91), which moves genetic information within the cell.

DNA is transcribed into other RNA molecules, such as those involved in splicing, transferring amino acids, and building ribosomes (see pp.92–93).

mRNA is translated from code in the form of bases on nucleic acid into an amino acid sequence.

THE CENTRAL DOGMA

Both DNA and proteins are molecules made up of long chains of subunits – bases for DNA, amino acids for proteins. The sequence of the subunits can be thought of as a string of digital information that carries data in a similar way to letters in a word or the binary code of computer language. DNA base sequences code for amino acid sequences in proteins, which – in turn – dictate protein shapes. These shapes determine what the protein molecules do in the body. This information flow is so fundamental to life that is has become known as biology's central dogma.

THREE-LETTER WORDS

A sequence of bases on a DNA molecule codes for a sequence of amino acids in a protein molecule. DNA has only four different kinds of bases, while proteins have 20 different amino-acid building blocks. This means that the genetic code must be more complex than a one-to-one correspondence between bases and amino acids. In fact, it is a sequence of three bases – a triplet – that codes for an amino acid. There are 64 possible combinations of three bases (for example, CGA and AAU), which is more than enough to code for 20 amino acids.

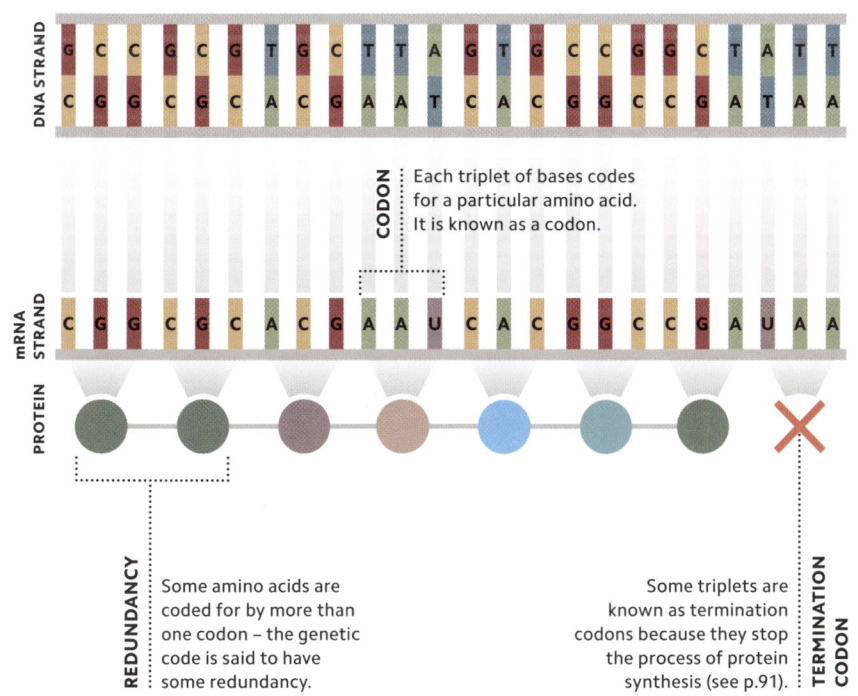

CODON Each triplet of bases codes for a particular amino acid. It is known as a codon.

REDUNDANCY Some amino acids are coded for by more than one codon – the genetic code is said to have some redundancy.

TERMINATION CODON Some triplets are known as termination codons because they stop the process of protein synthesis (see p.91).

TRIPLET CODE | 89

READING DNA

Genes are sections of DNA that code for specific proteins. They are contained in chromosomes in the cell nucleus. Proteins are assembled on ribosomes – structures outside the nucleus in the cell's cytoplasm. Genetic information must be moved from the nucleus to the ribosomes; it moves in form of a nucleic acid called messenger RNA, or mRNA (see p.63). Genes exposed by unwinding sections of the DNA double helix act as templates for making complementary mRNA strands in a process called transcription. The messengers – effectively mirror copies of genes – then move to ribosomes while the DNA molecule winds back up.

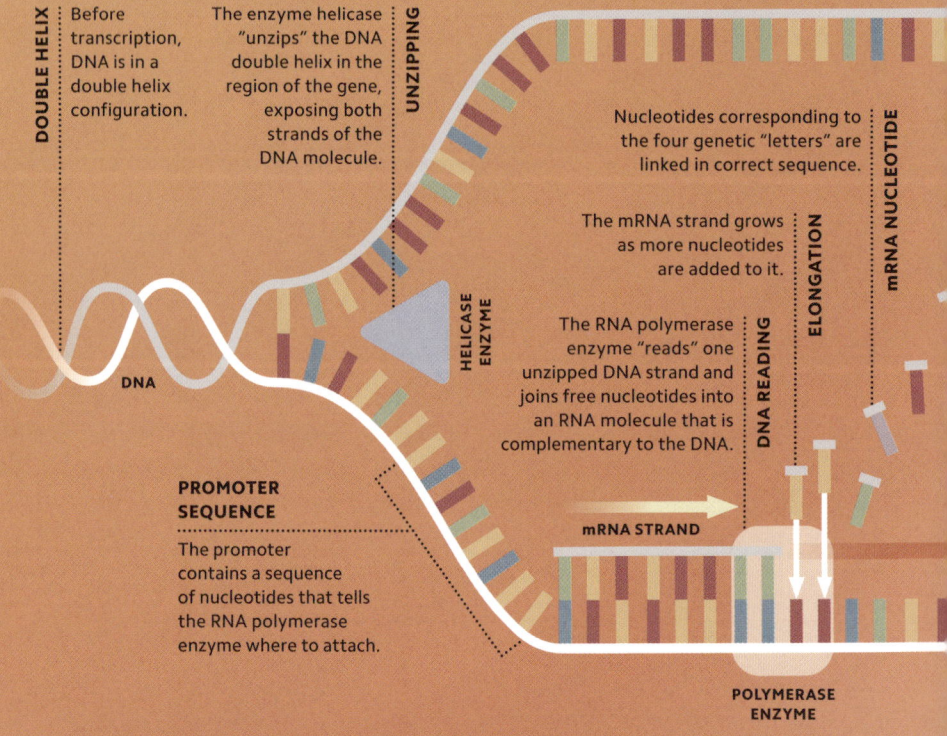

DOUBLE HELIX
Before transcription, DNA is in a double helix configuration.

UNZIPPING
The enzyme helicase "unzips" the DNA double helix in the region of the gene, exposing both strands of the DNA molecule.

PROMOTER SEQUENCE
The promoter contains a sequence of nucleotides that tells the RNA polymerase enzyme where to attach.

DNA READING
The RNA polymerase enzyme "reads" one unzipped DNA strand and joins free nucleotides into an RNA molecule that is complementary to the DNA.

ELONGATION
The mRNA strand grows as more nucleotides are added to it.

mRNA NUCLEOTIDE
Nucleotides corresponding to the four genetic "letters" are linked in correct sequence.

Transcription

When a gene is switched on (see pp.96–97), the DNA section corresponding to that gene "unzips", allowing enzymes to copy the gene into a strand of mRNA.

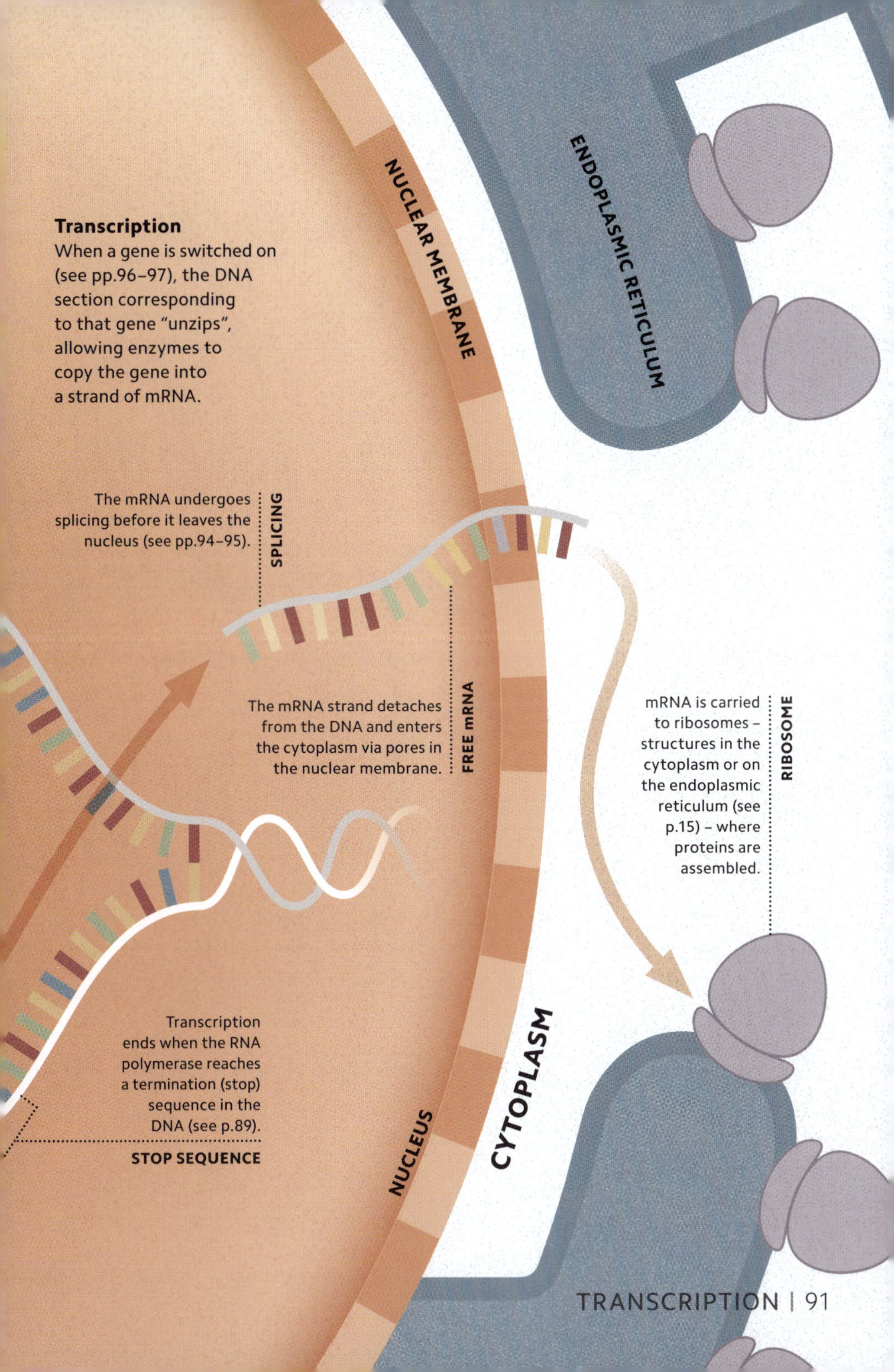

NUCLEAR MEMBRANE

ENDOPLASMIC RETICULUM

The mRNA undergoes splicing before it leaves the nucleus (see pp.94–95).

SPLICING

The mRNA strand detaches from the DNA and enters the cytoplasm via pores in the nuclear membrane.

FREE mRNA

mRNA is carried to ribosomes – structures in the cytoplasm or on the endoplasmic reticulum (see p.15) – where proteins are assembled.

RIBOSOME

Transcription ends when the RNA polymerase reaches a termination (stop) sequence in the DNA (see p.89).

STOP SEQUENCE

NUCLEUS

CYTOPLASM

TRANSCRIPTION | 91

BUILDING PROTEINS

Once bound to a ribosome, mRNA interacts with a complex of other molecules to link amino acids together into a protein. The sequence of amino acids is determined by the codons on the mRNA (see p.89). Ribosomes, which are made up of two subunits, grasp the mRNA molecule and move along its length reading each three-letter codon in turn. The ribosome docks to molecules of tRNA (see p.63) bringing amino acids to add to the growing protein chain. Once completed, the protein chain detaches and folds into the shape determined by its amino acid sequence.

Translation
In this process, instructions carried in mRNA molecules are translated into sequences of amino acids that make up proteins.

NUCLEAR EXIT After editing (see pp.94–95), a strand of mRNA leaves the nucleus via a pore in the nuclear membrane.

CODON Sets of three adjacent bases code for one amino acid.

92 | TRANSLATION

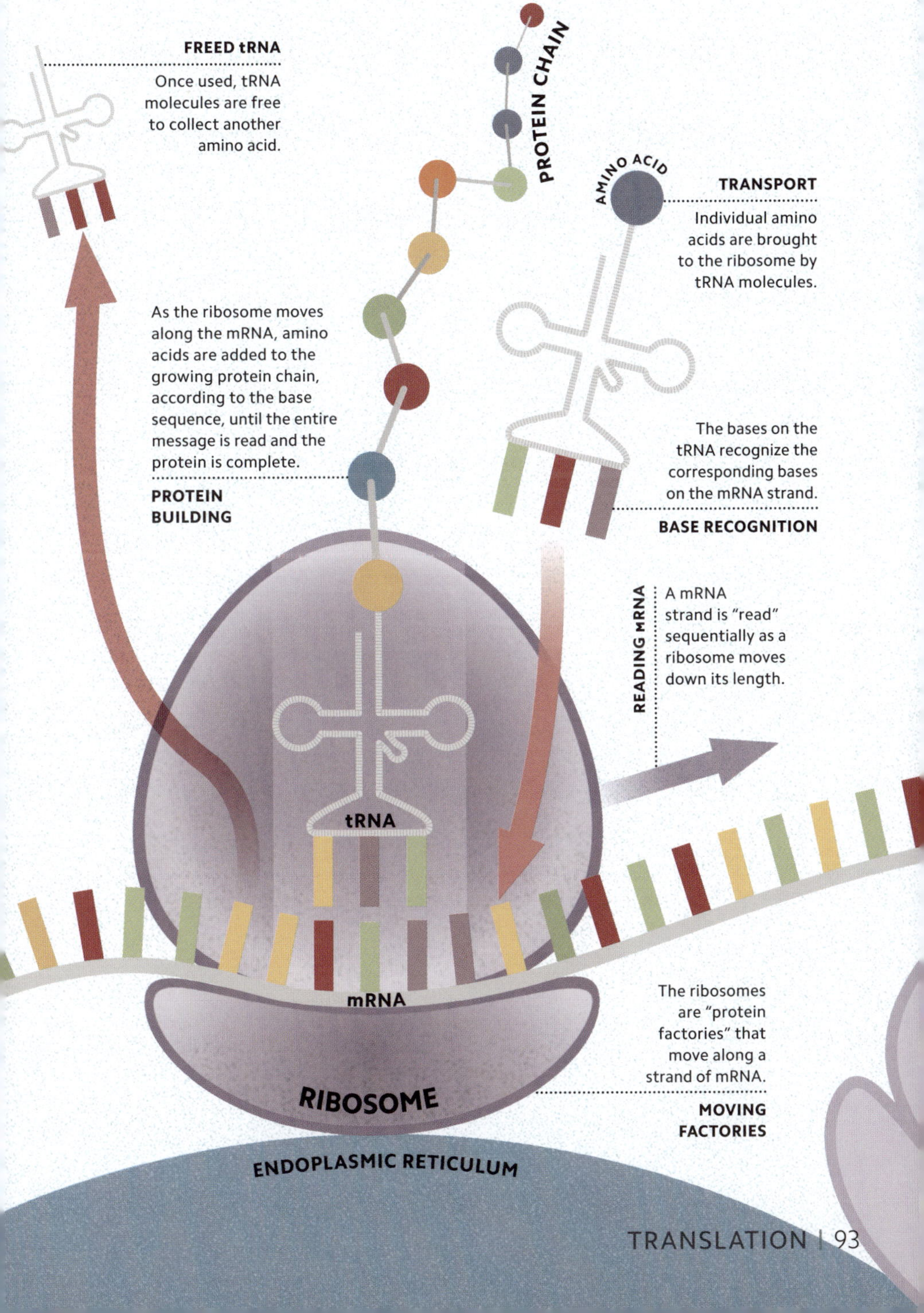

FREED tRNA

Once used, tRNA molecules are free to collect another amino acid.

As the ribosome moves along the mRNA, amino acids are added to the growing protein chain, according to the base sequence, until the entire message is read and the protein is complete.

PROTEIN BUILDING

PROTEIN CHAIN

AMINO ACID

TRANSPORT

Individual amino acids are brought to the ribosome by tRNA molecules.

The bases on the tRNA recognize the corresponding bases on the mRNA strand.

BASE RECOGNITION

READING mRNA

A mRNA strand is "read" sequentially as a ribosome moves down its length.

tRNA

mRNA

RIBOSOME

ENDOPLASMIC RETICULUM

The ribosomes are "protein factories" that move along a strand of mRNA.

MOVING FACTORIES

TRANSLATION

MAKING THE CUT

Plant and animal genes contain sections of DNA that do not code for amino acid sequences. These non-coding regions, called introns, punctuate the coding regions, or exons, and range in length from tens to tens of thousands of base pairs. Introns were once thought to be "junk" DNA (see p.64) but in fact have important roles: they may provide some protection against mutation, and also allow for "alternative splicing" – increasing the number of proteins a cell can produce. When RNA is transcribed from DNA, both introns and exons are transcribed together. The non-coding introns must be removed from messenger RNA before it can be used to make protein; this process is called RNA editing.

TRANSCRIPTION

EXON
These portions of DNA code for sequences of a protein.

INTRON
These portions of DNA do not code for sequences of a protein.

"START" MARKER SEQUENCE
These DNA sequences mark the start of an exon.

"END" MARKER SEQUENCE
These DNA sequences mark the end of an exon.

DNA STRAND

ON AND OFF SWITCHES

Not all the genes within a cell are active at the same time. Different proteins are needed at different times; for example, an enzyme that digests a sugar (such as lactose) is needed only when that sugar is available. Gene activity is regulated by substances in the cell that can either boost or depress transcription. They work by binding directly to the DNA. Some switch on genes by initiating the formation of mRNA; others block the manufacture of mRNA and therefore protein. Some alter the DNA-histone packaging to keep genes hidden, or expose them for transcription (see pp.90–91).

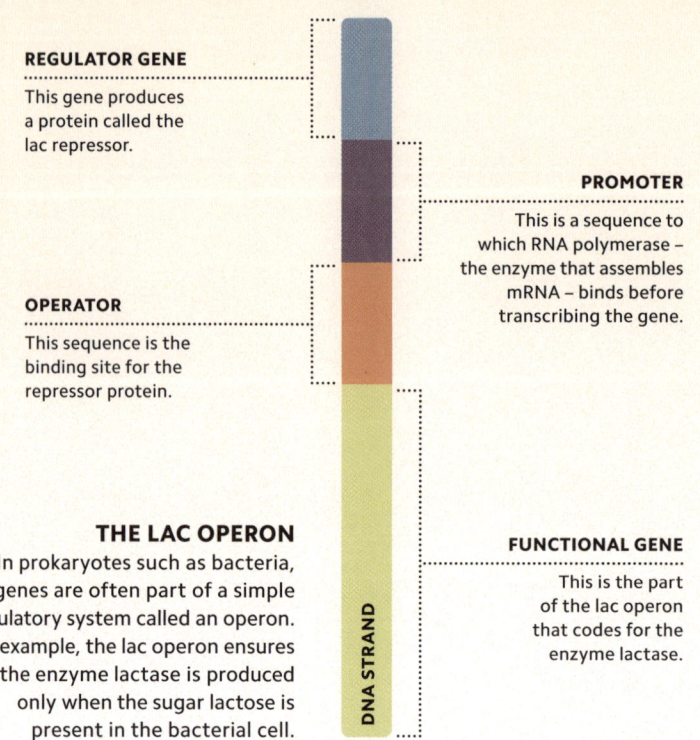

REGULATOR GENE
This gene produces a protein called the lac repressor.

PROMOTER
This is a sequence to which RNA polymerase – the enzyme that assembles mRNA – binds before transcribing the gene.

OPERATOR
This sequence is the binding site for the repressor protein.

THE LAC OPERON
In prokaryotes such as bacteria, genes are often part of a simple regulatory system called an operon. For example, the lac operon ensures that the enzyme lactase is produced only when the sugar lactose is present in the bacterial cell.

FUNCTIONAL GENE
This is the part of the lac operon that codes for the enzyme lactase.

DNA STRAND

LACTOSE ABSENT

If the sugar lactose is not available, the lac repressor protein binds to the operator, preventing the gene from being transcribed by the polymerase enzyme.

LAC REPRESSOR PROTEIN

RNA POLYMERASE

RNA polymerase binds to the promoter and tries to move along the gene.

BLOCKING TRANSCRIPTION

Lac repressor protein binds to the operator, so blocking the progress of the polymerase enzyme.

DNA STRAND

LACTOSE PRESENT

If lactose is available, it bonds to the repressor protein in such a way that the repressor can no longer attach to the operator. This switches the gene on.

REPRESSOR/LACTOSE COMPLEX

Lac repressor protein binds to lactose molecules and so cannot bind to the operator.

LACTOSE (SUGAR)

POLYMERASE

RNA polymerase is able to transcribe the gene to produce the lactase enzymes.

LACTASE (ENZYME)

LACTOSE (SUGAR)

LACTOSE SPLIT

Lactose is split and becomes available to the cell.

GENE REGULATION

CONTROL SYSTEMS

Almost every cell in the body of a eukaryotic organism carries the same genes, but different cell types – such as brain and blood cells – vary in structure and function. This is because different genes are switched on or off during their development and also in their day-to-day function. While genes in prokaryotes, such as bacteria, are regulated within operons (see pp.96–97), genes in eukaryotes tend to be regulated by more sophisticated mechanisms involving transcription factors.

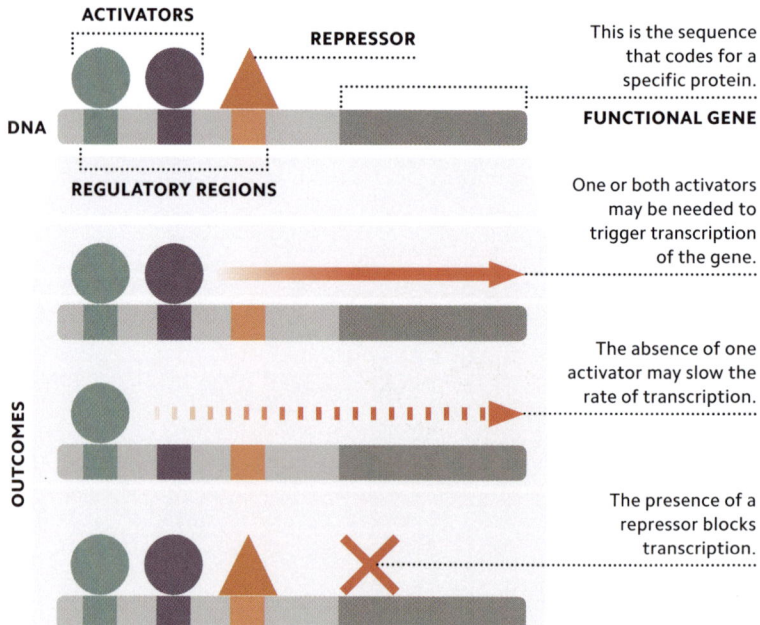

Transcription factors

Signals from inside or outside a cell activate proteins called transcription factors. These bind to regulatory regions of a gene. Some transcription factors are activators, increasing the rate at which the gene is transcribed into mRNA; others are repressors, blocking transcription of the gene.

SIGNAL

The signalling molecule may be a hormone in the bloodstream.

The hormone binds to a receptor on the surface of a target cell.

RECEPTOR

TRANSDUCTION

The binding of the hormone triggers the production of intermediary relay molecules, which trigger production of an activator protein.

RESPONSE

The activator protein binds to a regulatory region, switching on a specific gene, perhaps in combination with other activators.

SIGNALLING MOLECULE

RECEPTOR

CELL MEMBRANE

RELAY MOLECULES

ACTIVATOR PROTEIN

NUCLEAR MEMBRANE

TRANSCRIPTION

DNA

Cell signals

The signals that switch genes on and off may come from within the cell, and can be composed of many interacting factors. They may also come from outside the cell – hormones, for example, control gene activity.

GENES IN DEVELOPMENT

GENE TAGGING

Genes can be permanently or semi-permanently switched on or off during development in a process called epigenetic programming. This involves the chemical "tagging" of genetic material that make sections of DNA more or less readable to RNA polymerase enzymes (see p.90). These tags, or epigenetic factors, can attach to histone proteins around which DNA is wound (see p.62) or bind directly to DNA strands.

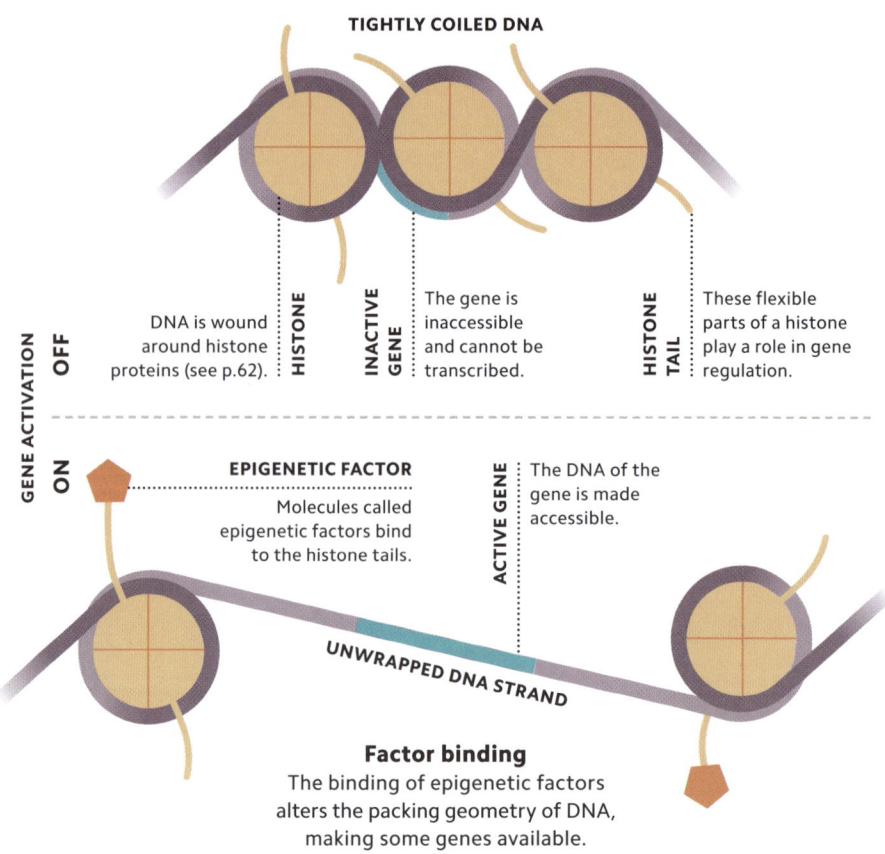

TIGHTLY COILED DNA

GENE ACTIVATION — OFF

- **DNA is wound around histone proteins (see p.62).**
- **HISTONE**
- **INACTIVE GENE**: The gene is inaccessible and cannot be transcribed.
- **HISTONE TAIL**: These flexible parts of a histone play a role in gene regulation.

GENE ACTIVATION — ON

- **EPIGENETIC FACTOR**: Molecules called epigenetic factors bind to the histone tails.
- **ACTIVE GENE**: The DNA of the gene is made accessible.

UNWRAPPED DNA STRAND

Factor binding
The binding of epigenetic factors alters the packing geometry of DNA, making some genes available.

CHEMICAL ASYMMETRY

The fates of cells are set early in their development by a series of chemical triggers. Some come into play during the first asymmetrical division of the fertilized egg (where one side of the egg has more yolk, while the other side has more active organelles). Others come from maternal proteins derived from cells surrounding the ovary. The chemical gradients that the developing embryo encounters cause its genetically identical cells to become specialized to make tissues and organs.

Cell fates
The early generations of cells in an embryo can develop into any type of cell (they are said to be totipotent). Exposure to chemical gradients induces them to specialize in structure and function.

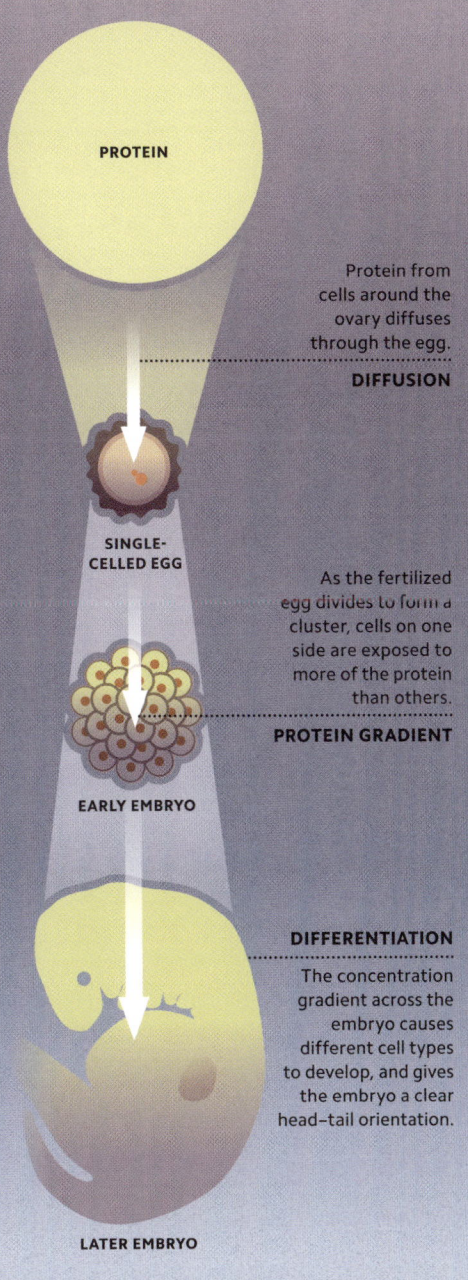

PROTEIN

DIFFUSION
Protein from cells around the ovary diffuses through the egg.

SINGLE-CELLED EGG

PROTEIN GRADIENT
As the fertilized egg divides to form a cluster, cells on one side are exposed to more of the protein than others.

EARLY EMBRYO

DIFFERENTIATION
The concentration gradient across the embryo causes different cell types to develop, and gives the embryo a clear head–tail orientation.

LATER EMBRYO

DRIVERS OF EARLY DEVELOPMENT

WHEN GO WRO

Genes play a fundamental role in life. Mutations – errors introduced into the DNA sequence of genes – can disrupt normal protein function, leading to conditions like cancer. Some mutations can be inherited and passed down through generations; these are genetic diseases, such as sickle cell. Mutations occur randomly, but certain factors, such as radiation exposure, can increase their incidence. On rare occasions, mutations have positive effects – for example increasing an individual's resilience to disease or tolerance of extreme temperatures. In this way, genetic diversity caused by mutations plays a crucial role in evolution.

RADIATION
Exposure to X-rays, high-energy particles, and ultraviolet light (present in sunlight) may cause DNA damage.

CHEMICALS
Some chemicals can cause DNA damage. Those that promote the development of cancer are called carcinogens.

MUTAGENIC AGENTS

Mutagenic agents
Some mutagenic agents occur naturally in the environment. Exposure to others can be reduced or avoided.

INFECTIONS
Infection by some viruses and bacteria may cause the production of chemicals that affect DNA structure.

COSTLY MISTAKES

Genetic variation results from sexual reproduction (see pp.74–75) but is also produced when mutations – changes to the structure of DNA – occur. Mutations happen naturally and by chance at random locations in an organism's genome. They may change the structure of a chromosome or disrupt the DNA molecule and its chemical building blocks. Some factors can significantly increase the rate of mutation. These are called mutagenic agents.

WHEN MUTATIONS OCCUR

Mutations that happen in the organs that produce sex cells (gametes) are copied into all cells of the offspring. This means the entire individual is affected by the mutation, and can also pass the mutation to its own offspring. However, if a mutation occurs in a "normal" (non-sex) cell in the body, it will only be copied to the cells descended from the mutated cell and cannot be passed on to later offspring.

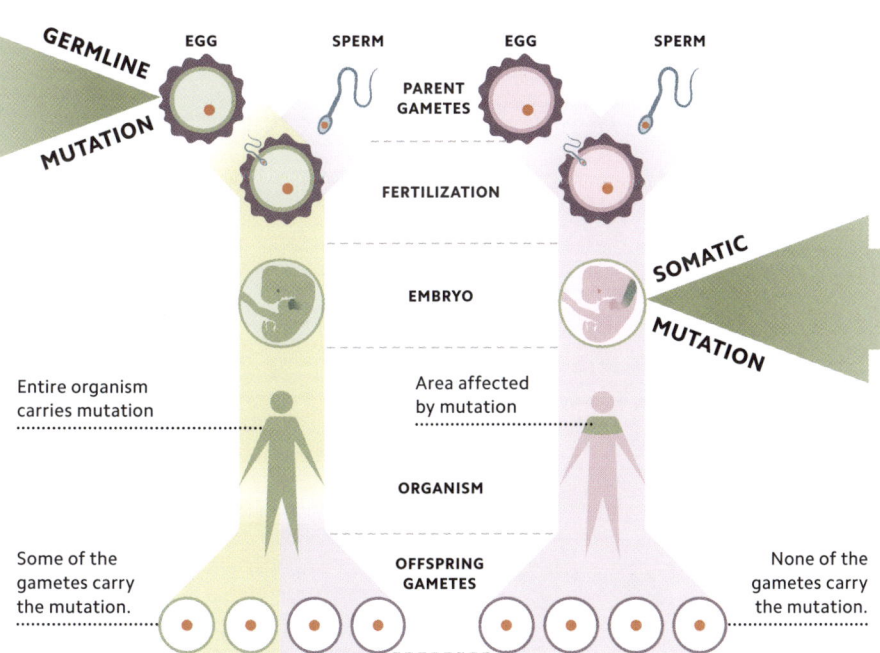

Germline mutation
A germline mutation occurs in the sex cells and is passed to all cells in the offspring. Some of the sex cells produced by the offspring carry the mutation.

Somatic mutation
A somatic mutation occurs after conception (in the embryo or any later stage of life). It can affect a large part of the offspring's tissue, but is not passed on.

COPYING ERRORS

A cell's mechanisms for DNA replication are remarkably accurate. DNA's structure itself (see pp.60–61) ensures that errors are rare, and "proofreading" enzymes detect and repair most remaining mistakes. However, with so much DNA in a cell, miscopying does still occur and may result in a gene mutation – a permanent change in the base sequence of a gene. The simplest type of gene mutation is when one base on the DNA molecule is mistakenly substituted with another.

Substitution mutation
A change of a single base on a DNA strand can cause the change of an amino acid, which may result in the production of a mutated protein. Single-base mutations can sometimes be serious (see pp.110–111).

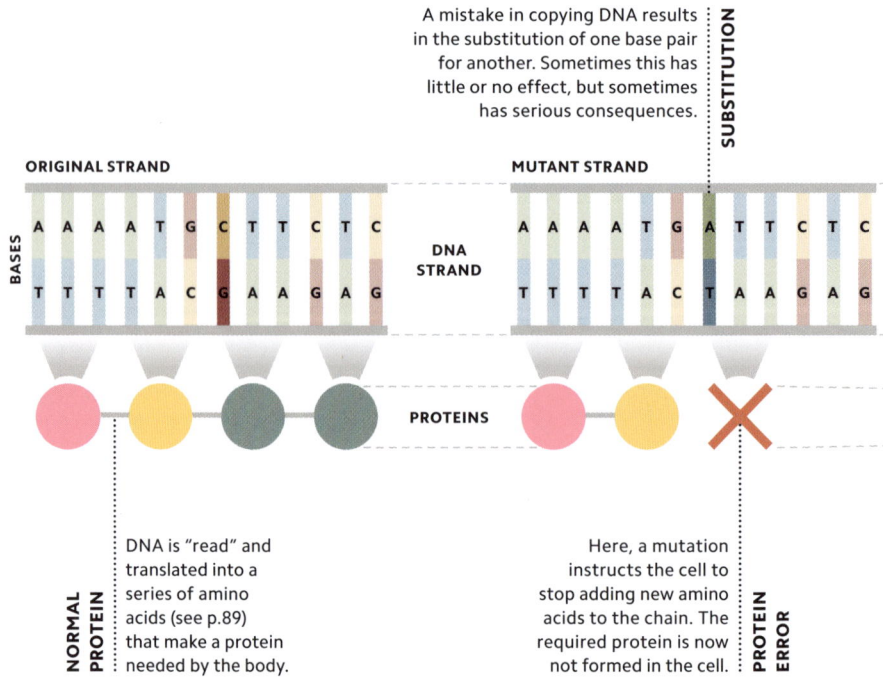

A mistake in copying DNA results in the substitution of one base pair for another. Sometimes this has little or no effect, but sometimes has serious consequences.

DNA is "read" and translated into a series of amino acids (see p.89) that make a protein needed by the body.

Here, a mutation instructs the cell to stop adding new amino acids to the chain. The required protein is now not formed in the cell.

106 | GENE MUTATIONS

SHIFT ALONG

Gene mutations in which a base is inserted into, or deleted from, a DNA strand are known as frameshift mutations, because the entire frame of reference for "reading" codons (triplets of bases on the DNA strand) is shifted. This results in multiple codons beyond the point of the mutation being misread. The protein resulting from this new sequence of bases is significantly altered from the normal state and is usually non-functional.

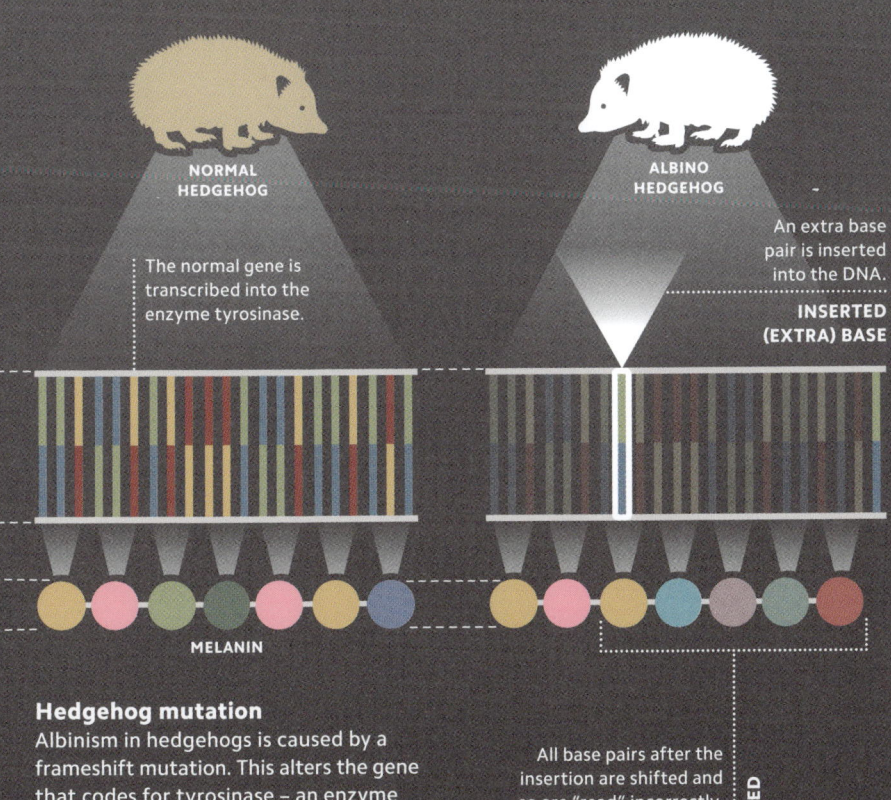

NORMAL HEDGEHOG

The normal gene is transcribed into the enzyme tyrosinase.

MELANIN

ALBINO HEDGEHOG

An extra base pair is inserted into the DNA.

INSERTED (EXTRA) BASE

All base pairs after the insertion are shifted and so are "read" incorrectly. No melanin is formed and the hedgehog is albino.

SHIFTED CODE

Hedgehog mutation
Albinism in hedgehogs is caused by a frameshift mutation. This alters the gene that codes for tyrosinase – an enzyme needed to make melanin, the pigment that gives colour to skin, hair, and eyes.

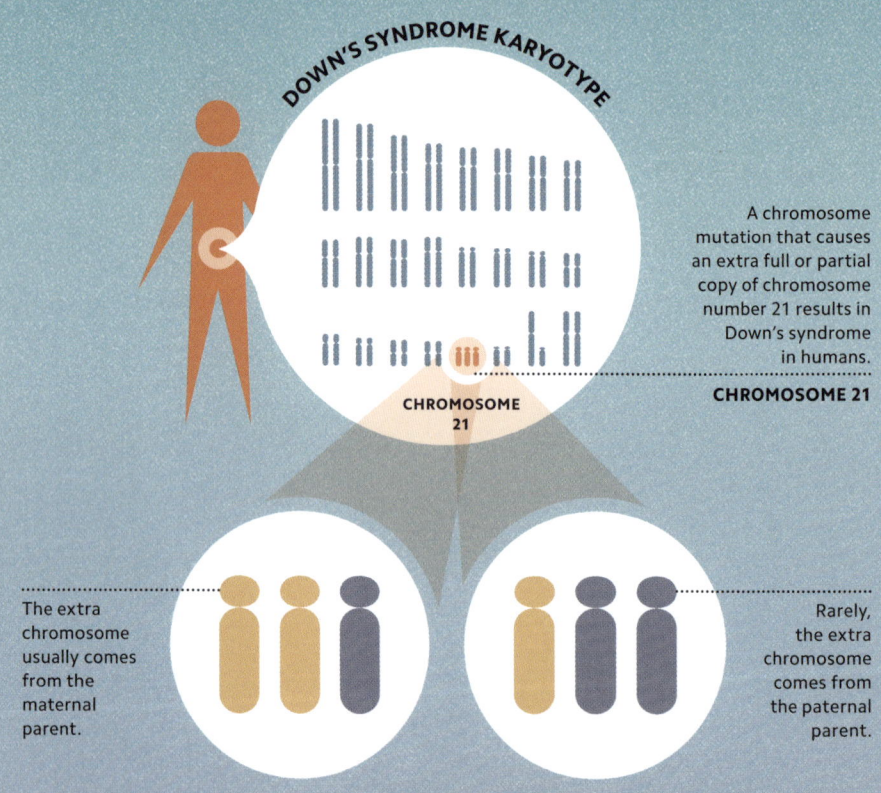

DOWN'S SYNDROME KARYOTYPE

CHROMOSOME 21

A chromosome mutation that causes an extra full or partial copy of chromosome number 21 results in Down's syndrome in humans.

CHROMOSOME 21

The extra chromosome usually comes from the maternal parent.

Rarely, the extra chromosome comes from the paternal parent.

TRISOMY OF CHROMOSOME 21

SORTING ERRORS

Genetic mutations may occur on a large scale, involving entire chromosomes rather than single genes. Such errors can arise when chromosomes fail to separate correctly during meiosis (see pp.78–81). If an individual has an extra copy of one chromosome, the result is called trisomy. Occasionally a whole extra set of chromosomes is inherited from either father or mother: the resulting embryo is called triploid, its cells carrying 69 rather than the normal 46 chromosomes. Triploidy usually ends in miscarriage or an early loss of a newborn.

MOVING SECTIONS

Many other types of chromosome mutation can result in serious, often lethal, developmental issues. These mutations result from errors in cell division that cause a section of a chromosome (spanning thousands to hundreds of thousands of bases) to break off, be duplicated, or erroneously become attached to another chromosome. There are four main ways in which these mutations can occur – duplication, deletion, inversion, and translocation.

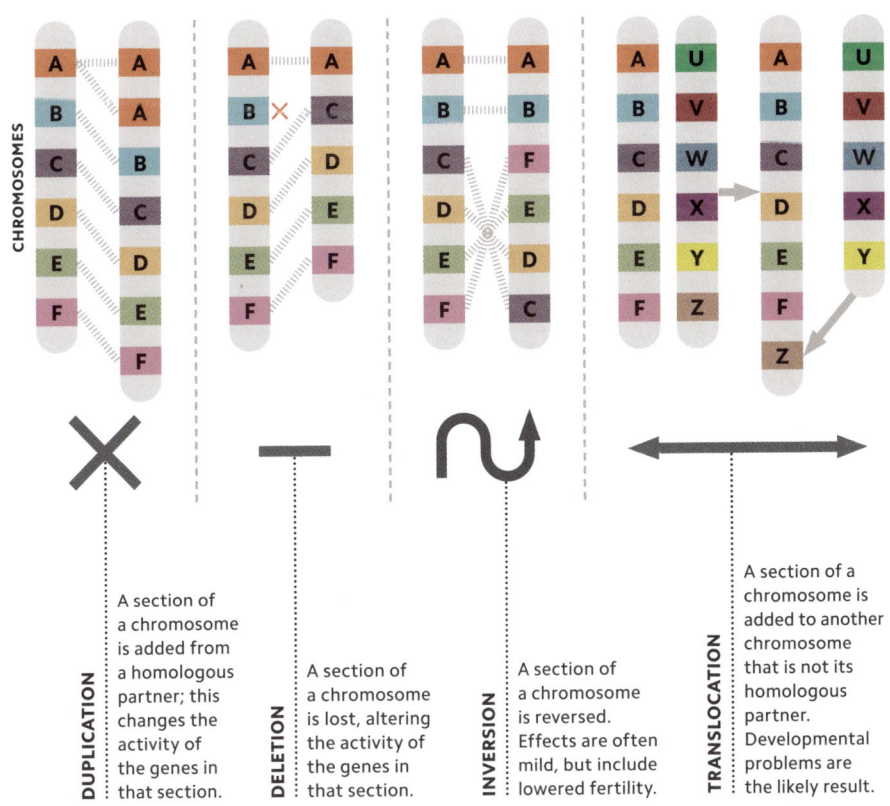

DUPLICATION: A section of a chromosome is added from a homologous partner; this changes the activity of the genes in that section.

DELETION: A section of a chromosome is lost, altering the activity of the genes in that section.

INVERSION: A section of a chromosome is reversed. Effects are often mild, but include lowered fertility.

TRANSLOCATION: A section of a chromosome is added to another chromosome that is not its homologous partner. Developmental problems are the likely result.

STRUCTURAL CHROMOSOME MUTATION | 109

RESHAPING PROTEINS

Genetic mutations change the sequence of bases along a DNA molecule. Triplets of bases that code for a specific amino acid (see p.89) may be disrupted by the mutation, leading to errors during protein synthesis. Sometimes these errors have no significant effect (because there is redundancy in the triplet code), but at other times they negatively affect the way the protein works, or stop it from being produced at all. Very rarely, they improve the performance of the protein.

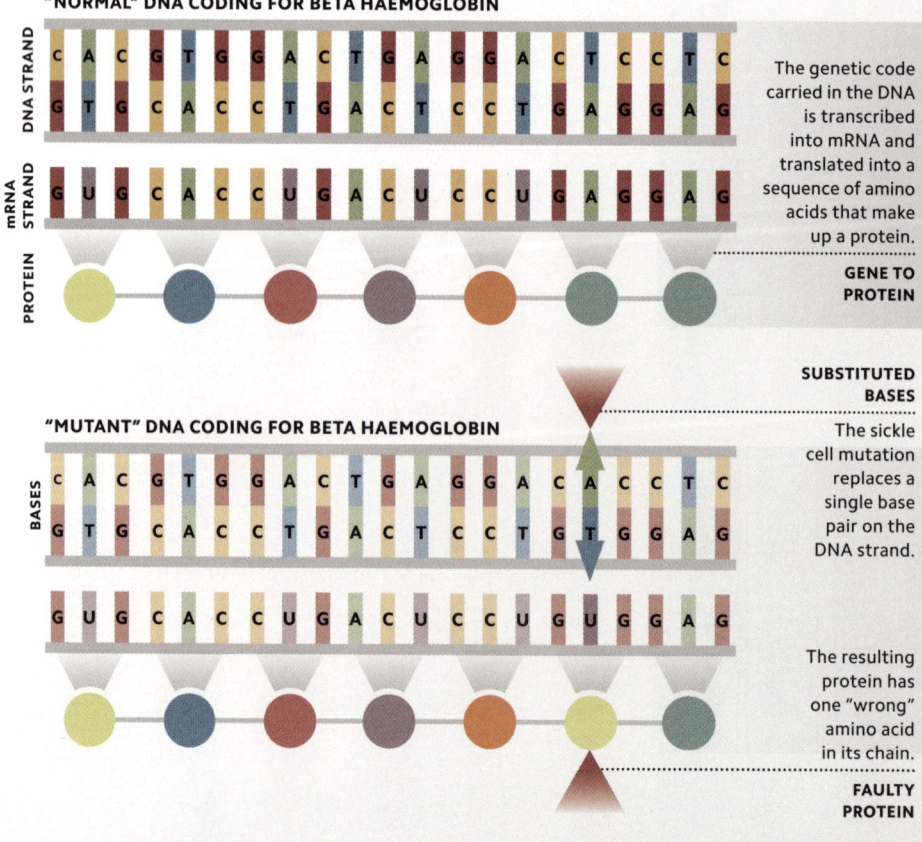

"NORMAL" DNA CODING FOR BETA HAEMOGLOBIN

The genetic code carried in the DNA is transcribed into mRNA and translated into a sequence of amino acids that make up a protein.

GENE TO PROTEIN

SUBSTITUTED BASES

"MUTANT" DNA CODING FOR BETA HAEMOGLOBIN

The sickle cell mutation replaces a single base pair on the DNA strand.

The resulting protein has one "wrong" amino acid in its chain.

FAULTY PROTEIN

Sickle cell disease

This genetic disease is caused by a single base substitution that affects the production of the protein beta haemoglobin – part of the haemoglobin complex that carries oxygen in red blood cells. Sufferers have unusually shaped red blood cells that may not live as long as healthy blood cells and can block blood vessels.

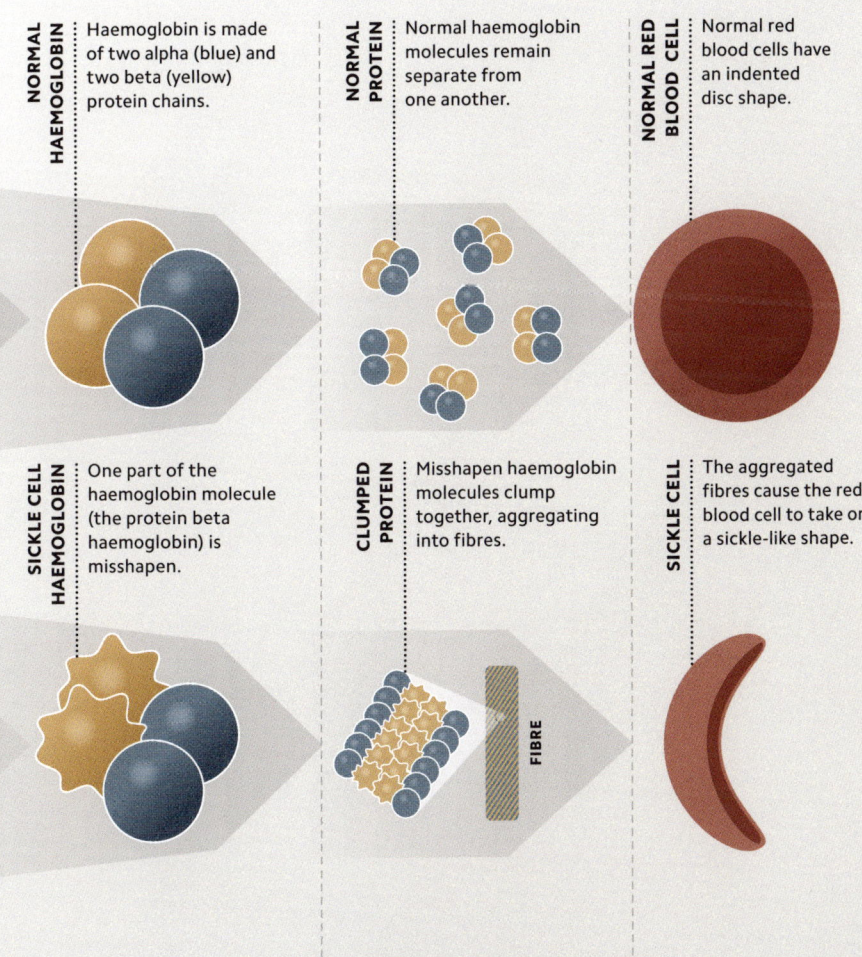

NORMAL HAEMOGLOBIN: Haemoglobin is made of two alpha (blue) and two beta (yellow) protein chains.

NORMAL PROTEIN: Normal haemoglobin molecules remain separate from one another.

NORMAL RED BLOOD CELL: Normal red blood cells have an indented disc shape.

SICKLE CELL HAEMOGLOBIN: One part of the haemoglobin molecule (the protein beta haemoglobin) is misshapen.

CLUMPED PROTEIN: Misshapen haemoglobin molecules clump together, aggregating into fibres.

SICKLE CELL: The aggregated fibres cause the red blood cell to take on a sickle-like shape.

UNWANTED LEGACY

Genetic diseases can result from changes in the structure or number of chromosomes or from mutations in genes (see pp.102–111). Inherited diseases that result from a change in just one gene are responsible for around 10,000 disorders in humans, including cystic fibrosis, Huntington's disease, and sickle cell disease. These disorders can easily be tracked through families and the risk of them occurring can be predicted. There are three groups of genetic diseases: dominant, recessive, and sex-linked.

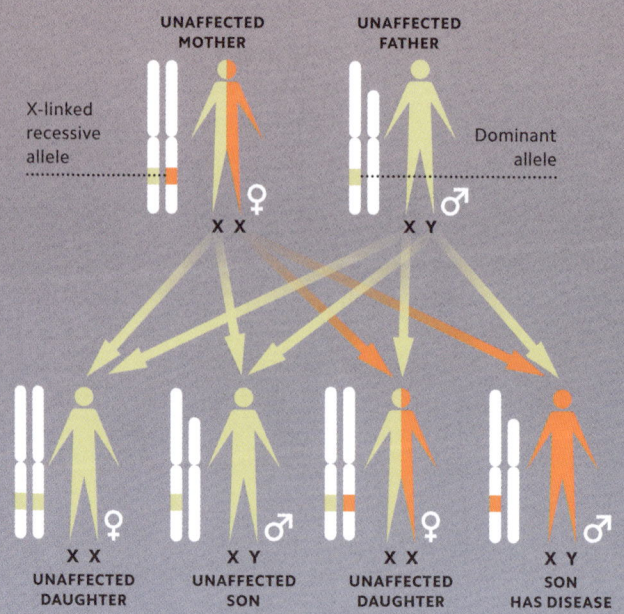

Sex-linked diseases
Diseases such as haemophilia are caused by a mutation in a gene on a sex (X or Y) chromosome. They are much more common in males because male offspring only have one X chromosome (see p.19) so recessive alleles are expressed if present. Only one copy of the mutant allele is required in males for the disease to occur.

Dominant diseases

Huntington's disease is caused by a dominant mutant allele. It tends to arise in every generation of an affected family. Because the mutant allele is dominant, everyone who carries it displays symptoms of the disease.

If one parent carries a Huntington's allele, children have a 50 per cent chance of being affected.

Recessive diseases

Cystic fibrosis (CF) is carried by a recessive mutant allele. Two copies of the faulty allele are required for a person to display symptoms. If a person only inherits one copy of the allele they are a "carrier", but are unaffected.

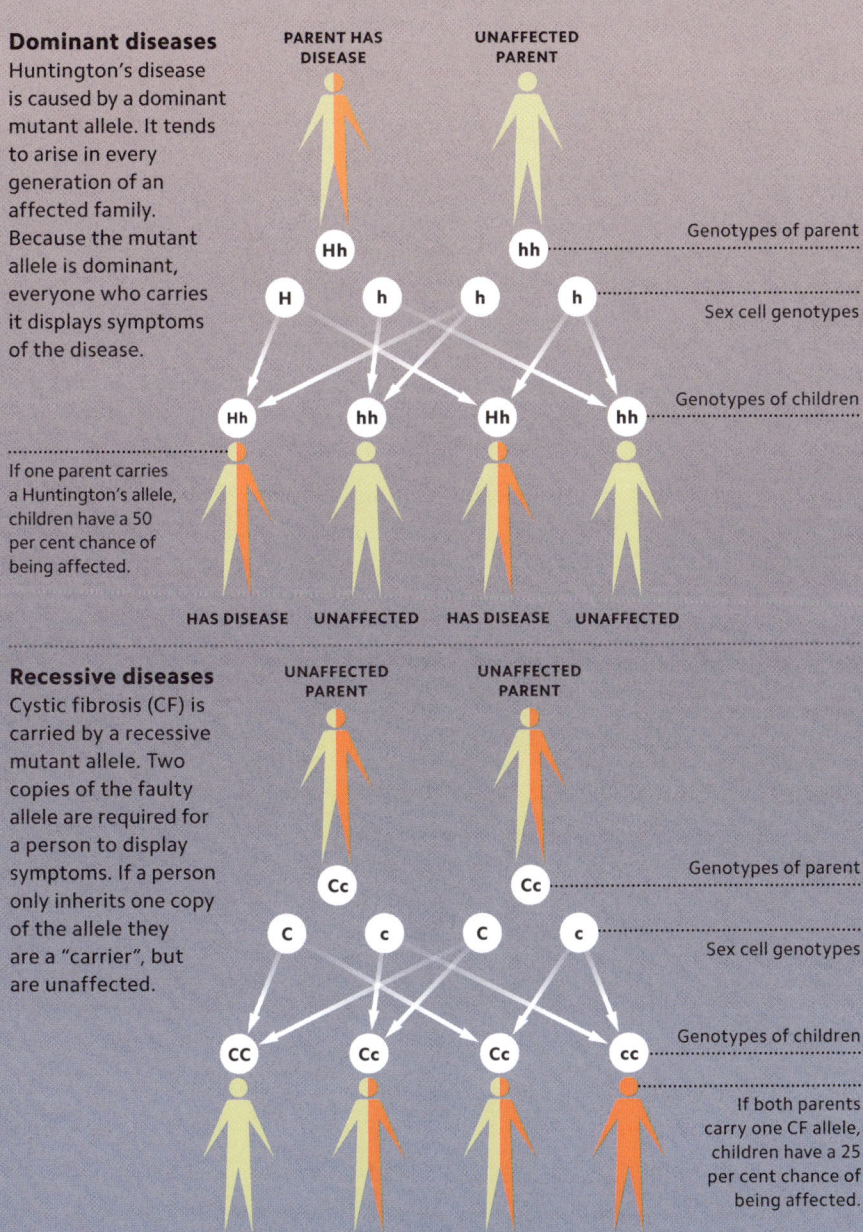

If both parents carry one CF allele, children have a 25 per cent chance of being affected.

GENETIC DISEASES

POPULA
GENETI
EVOLUT

The laws of inheritance ensure continuity from one generation to the next as individuals interbreed within their populations. However, change is an intrinsic part of the genetics story: new characteristics result from mutations and genetic "reshuffling" in sexual reproduction. As these changes accumulate over time, organisms evolve – and new species emerge while others become extinct. Indeed, all life is connected by a common ancestry that stretches back thousands of millions of years. Studying the distribution of genes in populations over space and time reveals much about the processes of natural selection and evolution.

Gene pools
The total set of genes and variations among those genes (alleles) within a population is known as its gene pool.

Different shapes represent different genes.

Different colours represent different alleles of a gene.

GENES IN POPULATIONS

Genes can be studied at the level of individuals – how they are transmitted from one generation to the next – and also at the level of populations. A population of a sexually reproducing organism is made up of interbreeding individuals of the same species. All these individuals possess the genes that define their species, but differ in the varieties of these genes, or alleles, as well as the combinations in which these alleles occur, or genotypes. The numbers and proportions of alleles provide a way of expressing the genetic make-up of the entire population (the gene pool).

KEEPING STABLE

Members of a population interbreed, so the combinations of alleles in the offspring differ from those in the parents according to the laws of inheritance. However, under unchanging conditions, the proportions of alleles within the population as a whole remain the same. Such a population is said to be in stable genetic equilibrium. This is the Hardy-Weinberg principle, named after the mathematician and physician who independently established the idea in the early 20th century.

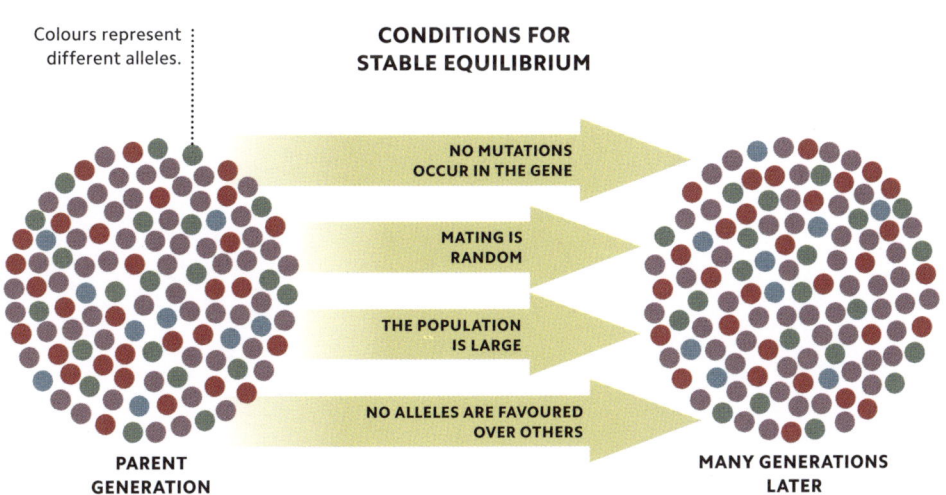

Hardy-Weinberg principle
Equilibrium is achieved and the number of alleles remains the same through generations of a population, but only if certain conditions (above) are met.

GENETIC EQUILIBRIUM

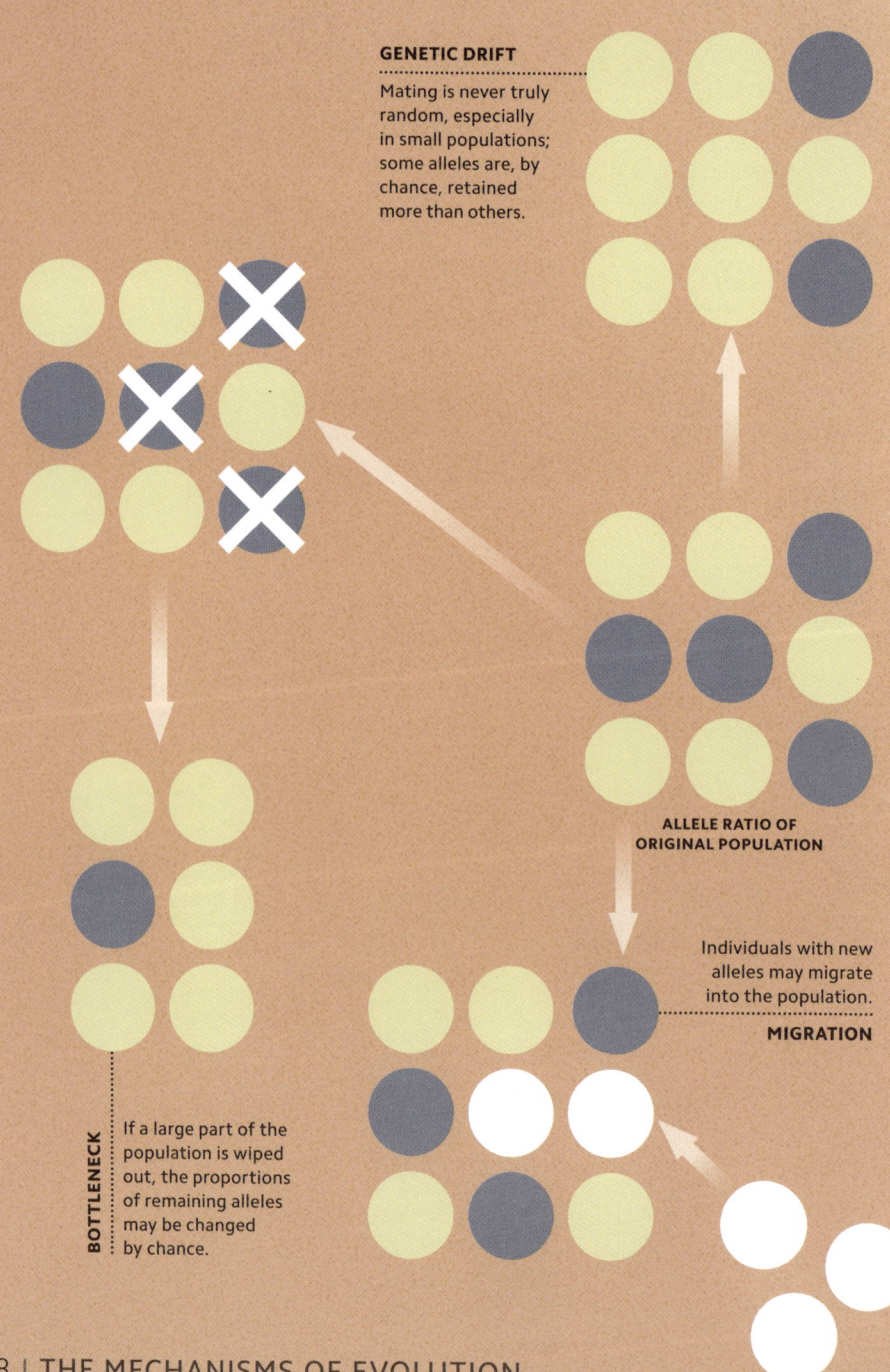

MUTATION
Random mutations in the DNA of individuals may create a new allele that substantially changes a characteristic.

Marking time
There are five main drivers of allele change, and thus evolution, in a population. They may work individually or in combination.

AGENTS OF CHANGE

In the natural world, populations rarely remain in stable equilibrium (see p.117) for long periods of time. A number of factors – some internal, such as mutation, and some external, such as changes in the environment – can and do affect the ratios of alleles within the population. The changes might be slight, but over centuries and hundreds of generations they change the characteristics of the population as a whole. And over millions of years, these accumulated changes can generate new species and diverging groups of plants and animals.

NATURAL SELECTION
If one allele confers a characteristic that makes the organism "fitter" and more likely to survive and reproduce, that allele becomes prevalent in the population.

THE MECHANISMS OF EVOLUTION

FORTUNATE MISTAKES

The cellular machinery responsible for DNA replication is precise, but not infallible. Over time, copying errors change genes. Many of these mistakes cause cells to malfunction; others have little or no effect or are survivable. A few (such as the lactase mutation) are beneficial and are passed to the offspring. The fact that all life on Earth has descended from a single microbial common ancestor points to the importance of mutation as a driver of evolution – it has generated the new genes that account for the rich diversity of plants and animals living today.

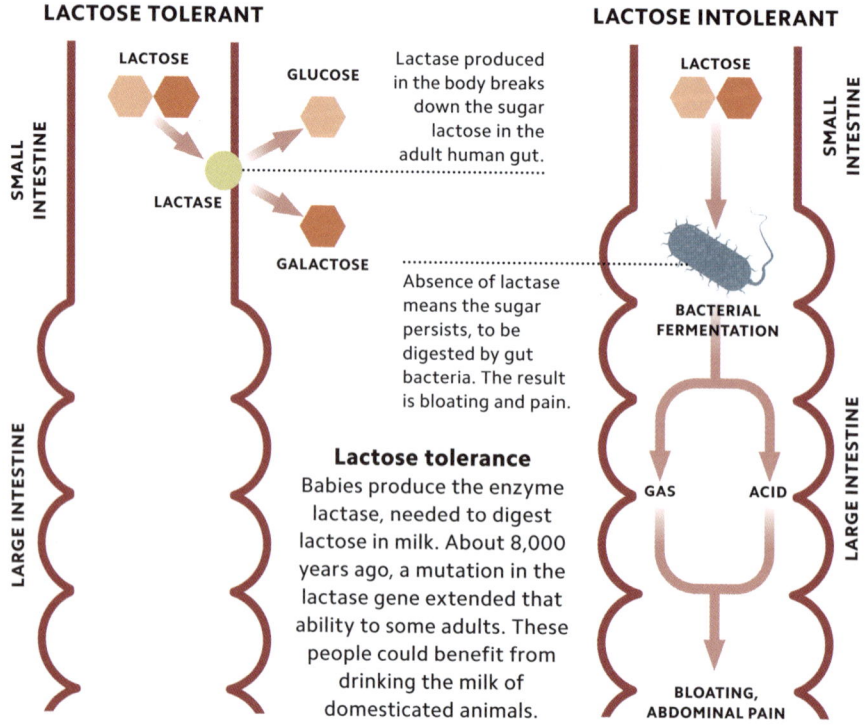

LACTOSE TOLERANT

Lactase produced in the body breaks down the sugar lactose in the adult human gut.

Absence of lactase means the sugar persists, to be digested by gut bacteria. The result is bloating and pain.

Lactose tolerance
Babies produce the enzyme lactase, needed to digest lactose in milk. About 8,000 years ago, a mutation in the lactase gene extended that ability to some adults. These people could benefit from drinking the milk of domesticated animals.

LACTOSE INTOLERANT

CHANCE CHANGE

Chance can play a big part in changing the proportion of alleles in a population, particularly if that population is small. For example, if half of a population is killed by a disease, it is likely that some alleles will be more numerous than others in the remaining individuals. Such a genetic "bottleneck" can reshape the population, as long as it remains isolated from the wider pool of genes. A similar phenomenon called the founder effect is often seen in island populations, which evolve more rapidly than larger mainland populations.

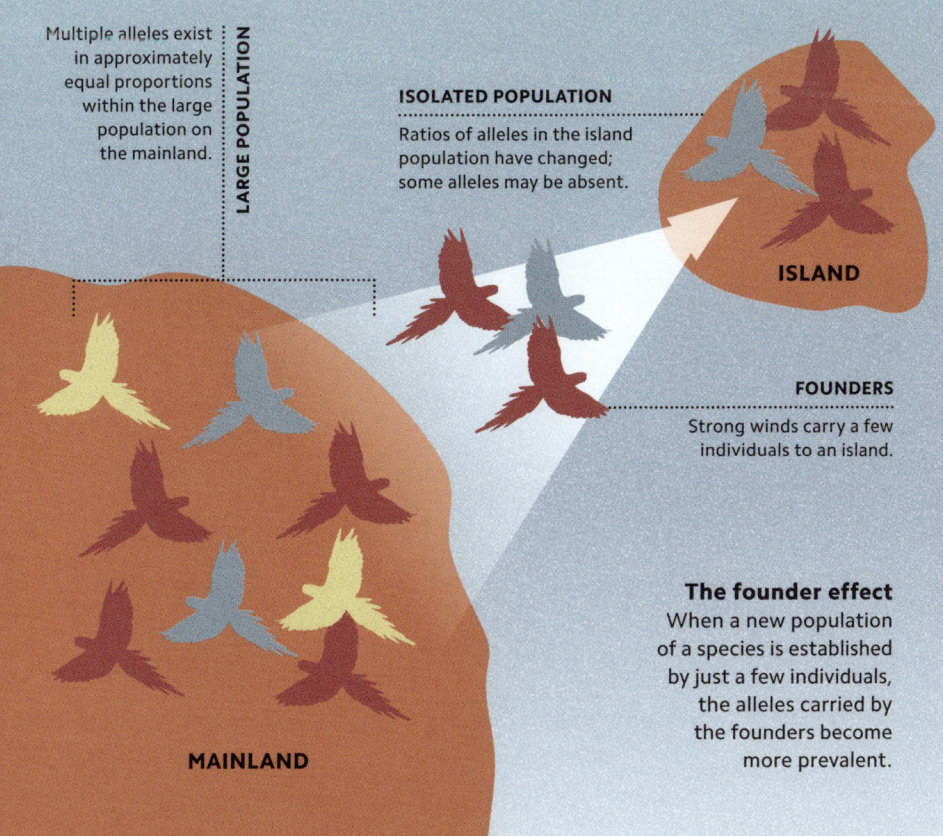

Multiple alleles exist in approximately equal proportions within the large population on the mainland.

LARGE POPULATION

ISOLATED POPULATION
Ratios of alleles in the island population have changed; some alleles may be absent.

ISLAND

FOUNDERS
Strong winds carry a few individuals to an island.

The founder effect
When a new population of a species is established by just a few individuals, the alleles carried by the founders become more prevalent.

MAINLAND

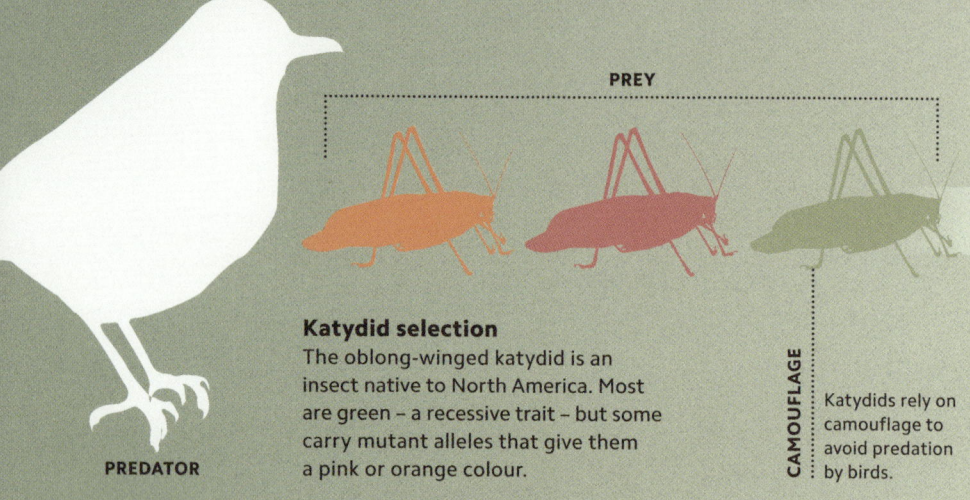

Katydid selection
The oblong-winged katydid is an insect native to North America. Most are green – a recessive trait – but some carry mutant alleles that give them a pink or orange colour.

PREDATOR

PREY

CAMOUFLAGE
Katydids rely on camouflage to avoid predation by birds.

BEST FIT

Genes govern an organism's characteristics, and so dictate how it interacts with its environment. Some characteristics (and therefore alleles) make an organism better suited to its environment, others worse suited. For example, a genetic change that reduces the density of a mammal's fur may be detrimental in a cold environment but beneficial in a hot one. The "fittest" individuals (those that possess characteristics that best fit the demands of their environment) are more likely to breed and pass their "fit" alleles on to the next generation. This process, known as natural selection, is one of the main drivers of evolution.

> The theory of natural selection was put forward by Charles Darwin in 1859 to account for the adaptations of living things.

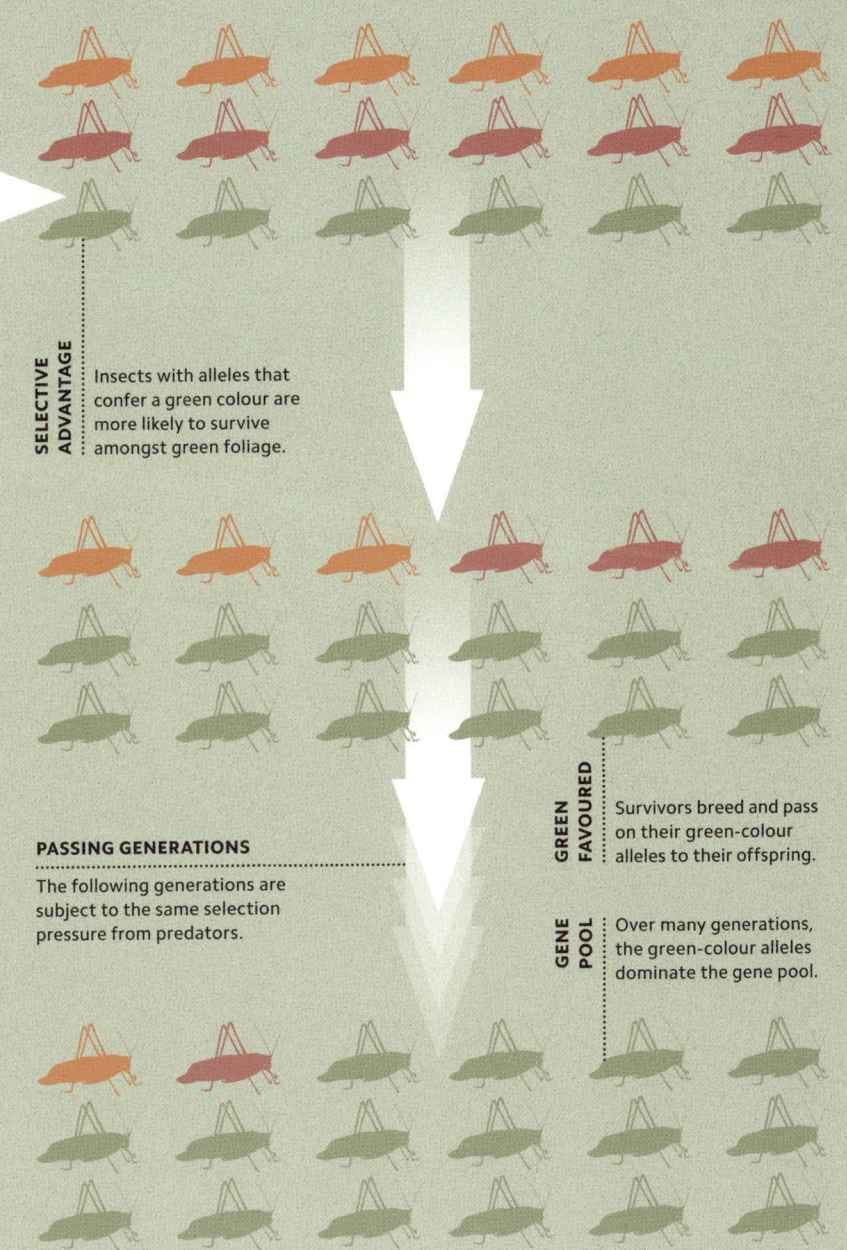

SELECTIVE ADVANTAGE: Insects with alleles that confer a green colour are more likely to survive amongst green foliage.

PASSING GENERATIONS: The following generations are subject to the same selection pressure from predators.

GREEN FAVOURED: Survivors breed and pass on their green-colour alleles to their offspring.

GENE POOL: Over many generations, the green-colour alleles dominate the gene pool.

POPULATION PRESSURES

By retaining characteristics best suited to the current environment and eliminating others, natural selection continually shapes the physical and behavioural traits of organisms. Selection can work in a number of ways. The most straightforward is directional selection, in which increasingly extreme versions of a particular characteristic are favoured. Selection can also stabilize characteristics, eliminating extremes and favouring an average state. Disruptive selection has the opposite effect, favouring more than one optimal state.

No selection
If natural selection neither favours or disfavours a trait, the distribution of that trait in the population will be symmetrical about an average value.

> More offspring are produced than can survive. Natural selection works to favour the survival of those best adapted to their environment.

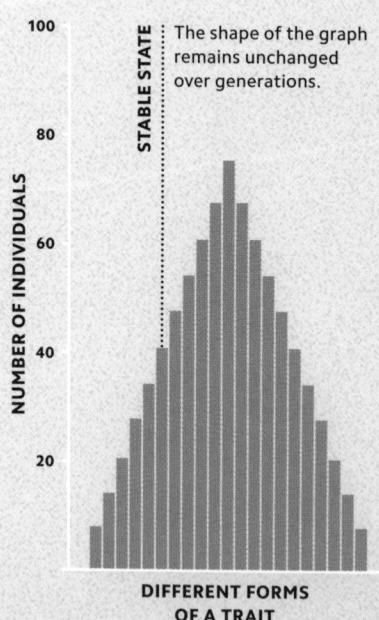

The shape of the graph remains unchanged over generations.

Directional selection

Selection that favours an extreme variant of a characteristic will move the entire distribution of that characteristic in one direction.

Giraffes with progressively longer necks are favoured.

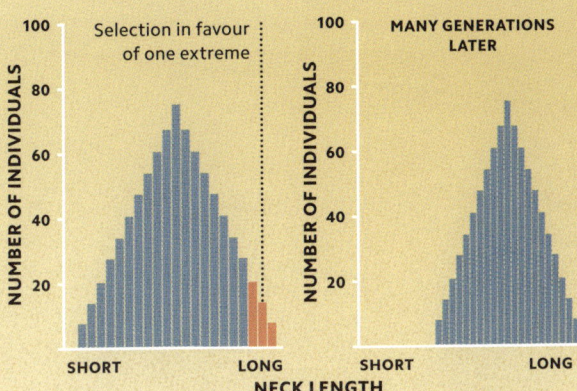

Stabilizing selection

Such selection brings the distribution of characteristics closer to the average.

Birth weight in mammals tends towards the average.

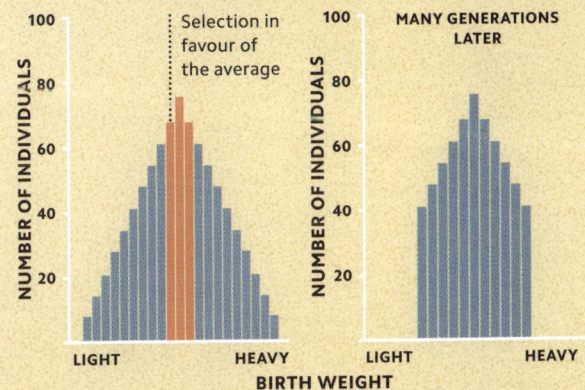

Disruptive selection

This selection favours the formation of two (or more) discrete types.

Light and dark shell forms are camouflaged but intermediate ones are taken by predators.

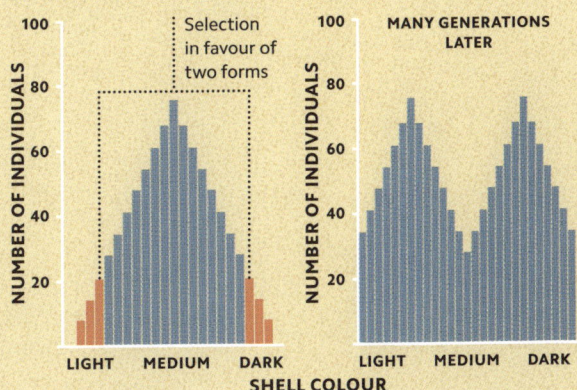

TYPES OF SELECTION | 125

Bird of paradise
A female bird will choose a mate with the largest, most vibrant feathers. Such selection amplifies these male attributes in subsequent generations, allowing them to find mates more easily.

THE MATING GAME

Natural selection causes various traits in populations to change over time. One type of selection – known as sexual selection – works on the reproductive process itself. Female animals usually invest more energy into the production of sex cells and the care of young than do males, so there is a great incentive for them to choose the healthiest, fittest mate available. Males signal their fitness by investing in the production of exaggerated characteristics, such as the large antlers of deer, or energetic courtship behaviour. Male bowerbirds, for example, build elaborate structures, or bowers, which they decorate to attract females.

SCALE IN EVOLUTION

Genetic mutations are capable of producing sudden, large-scale change, but the resulting organisms – so-called "hopeful monsters" – rarely survive to establish their own lineage. Evolutionary change is typically a gradual process of small increments – microevolution – that accumulate to produce larger effects that collectively constitute a new species. More often, species are formed when populations split into separate lineages that change, along different trajectories, into new species. With greater accumulated change, the trajectories can become entirely different taxonomic groups (macroevolution), as when reptilian ancestors evolved into birds and mammals.

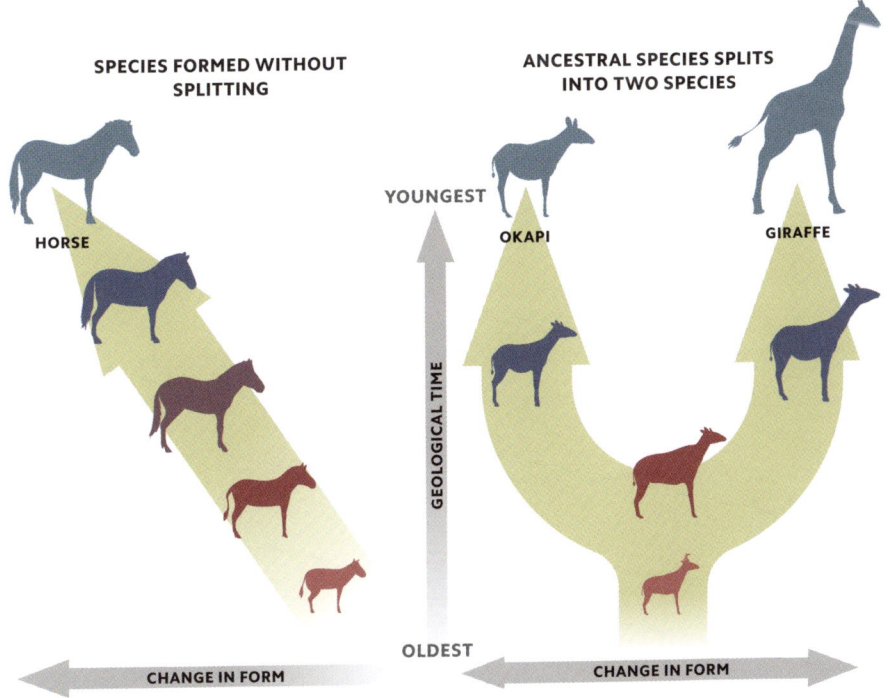

MICRO- AND MACROEVOLUTION | 127

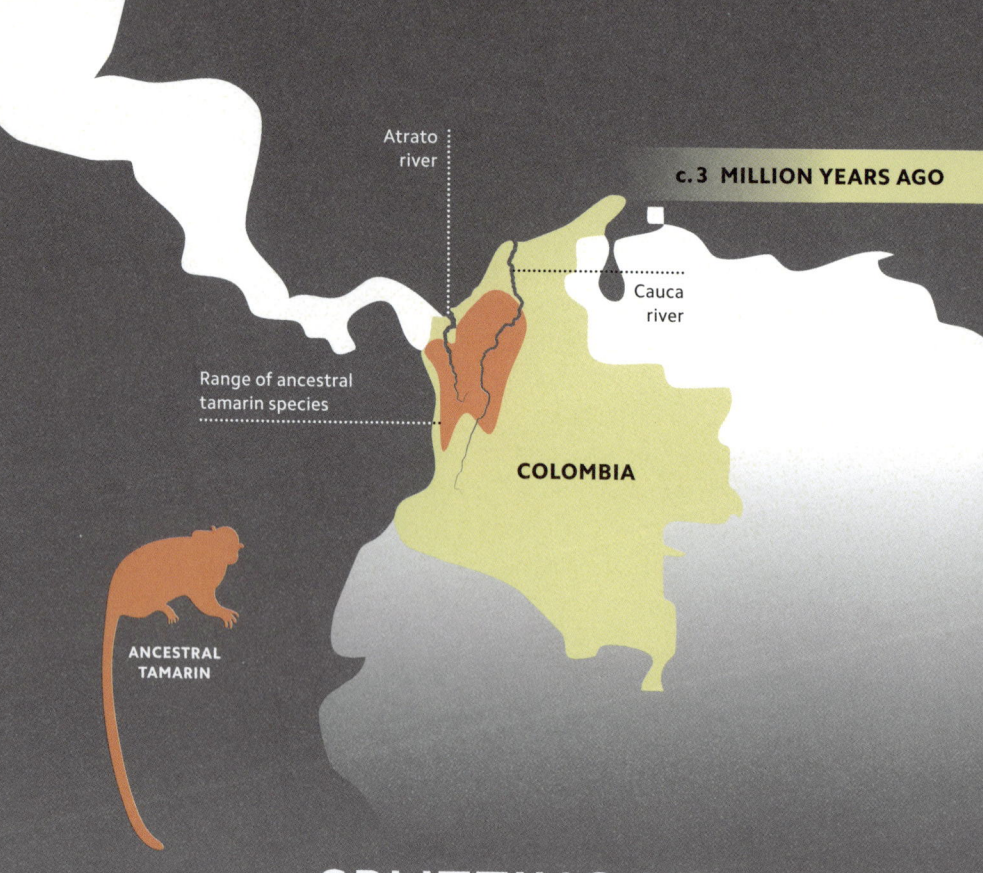

SPLITTING UP

Species can form when a population is split into two or more parts. A flood, a change in course of a river, or an earthquake may split up a population, or part of it may migrate away from the main group. Now isolated from one another, the populations experience different selection pressures (see pp.122–23) and so evolve along different paths. Over generations, they diverge so much that they cannot successfully interbreed with one another. At this point, the populations are considered to be different species. This type of species formation (or speciation) is called allopatric (from the Greek words for "other" and "fatherland") because it requires geographical separation.

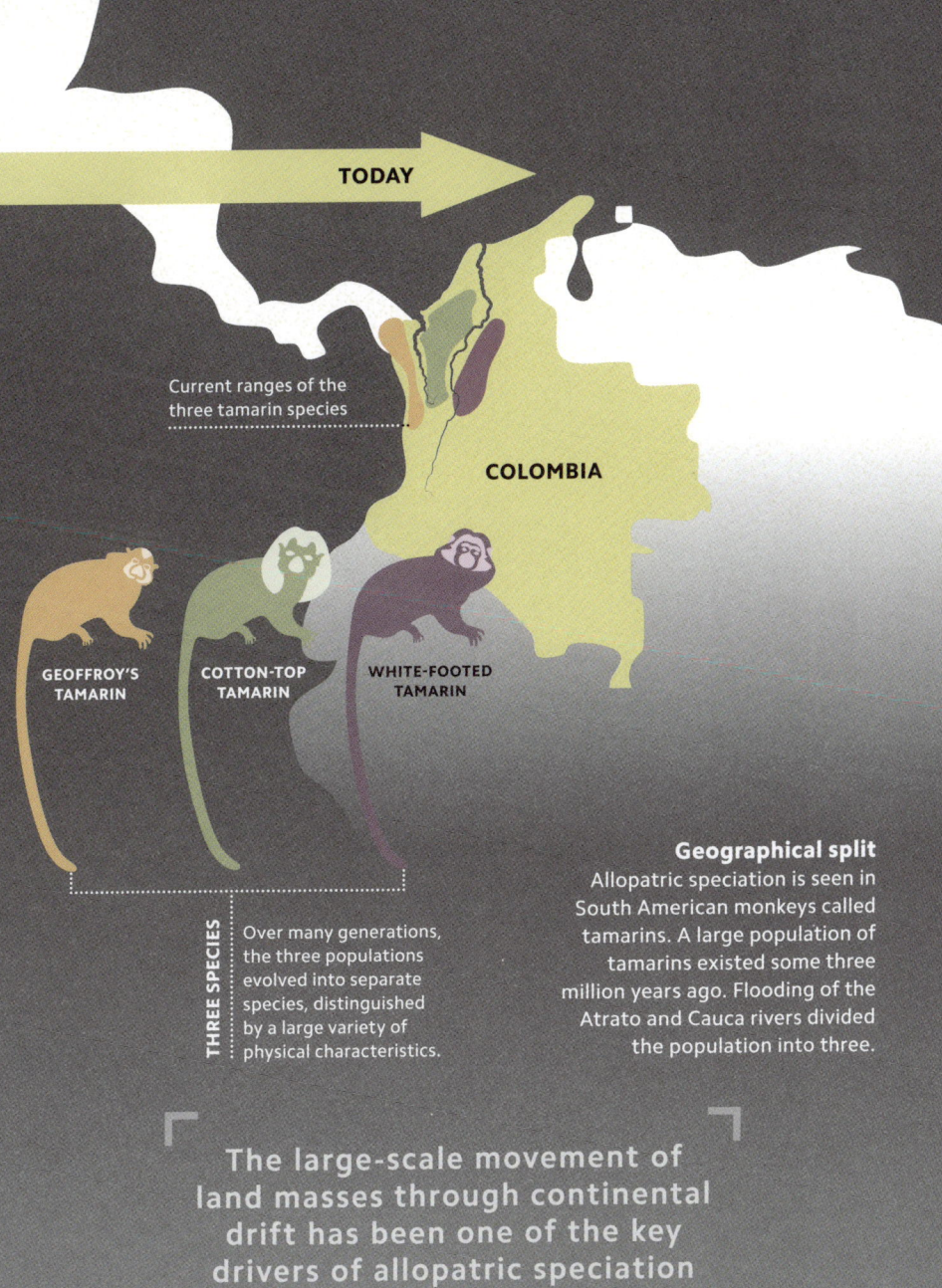

EVOLVING TOGETHER

It is possible for species to form without the need for geographical isolation (see pp.128–29). Quirks of cell division can produce sudden changes in an individual's chromosome number that prevent it from producing fertile offspring with its neighbours. These types of chromosome mutations are especially common in plants; around a third of plant species have multiple sets of chromosomes generated in this way. This phenomenon is called polyploidy. If enough individuals with a novel chromosome mutation interbreed, a new species can be created within a few generations. This form of speciation is called sympatric (from the Greek for "same fatherland").

Polyploidy
An error occurs as a plant's sex cells are formed. The chromosomes fail to separate during meiosis, meaning that the gametes are diploid rather than haploid (see pp.76–77).

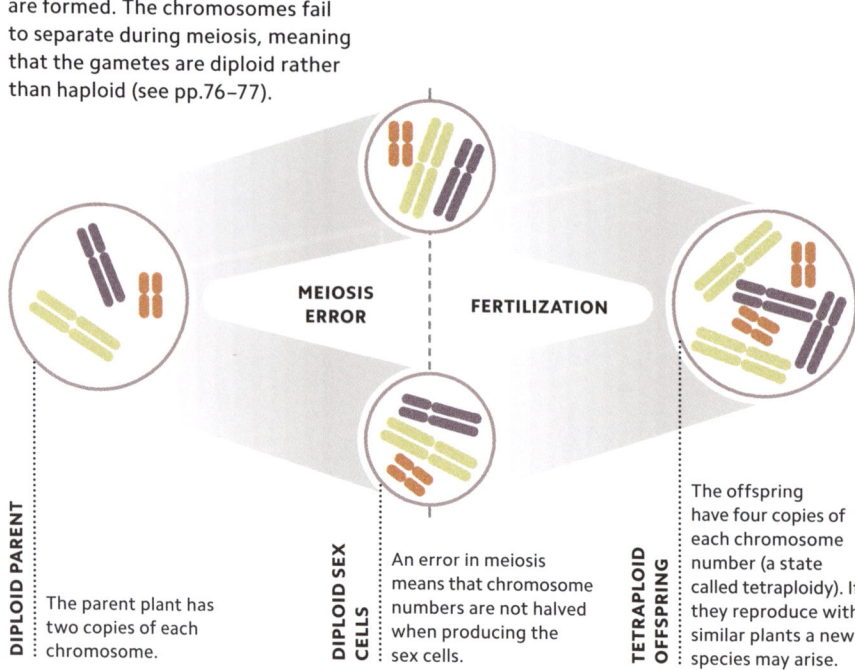

DIPLOID PARENT
The parent plant has two copies of each chromosome.

MEIOSIS ERROR

DIPLOID SEX CELLS
An error in meiosis means that chromosome numbers are not halved when producing the sex cells.

FERTILIZATION

TETRAPLOID OFFSPRING
The offspring have four copies of each chromosome number (a state called tetraploidy). If they reproduce with similar plants a new species may arise.

SYMPATRIC SPECIATION

KEEPING APART

TEMPORAL BREEDING BARRIER
Related species of frogs may reach sexual maturity at different times of year, so preventing interbreeding.

ECOLOGICAL BREEDING BARRIER
Species of garter snake occupy different habitats – one is terrestrial, another aquatic – preventing breeding between the species.

MECHANICAL BREEDING BARRIER
Related plant species may rely on different pollinating insects, so preventing crossbreeding between species.

The process of speciation may begin when a population is split into two separate parts (see pp.128–29). However, for the two populations to be considered separate species, a mechanism must evolve to prevent interbreeding should the two parts of the population be reunited. Some of these breeding barriers come into play before fertilization occurs, some after fertilization has happened.

BEHAVIOURAL BREEDING BARRIER
Birds will only respond to mating calls made by individuals of the same species. Even small differences in song can be a breeding barrier.

HYBRID DYSFUNCTION BREEDING BARRIER
Hybrids formed by the mating of two related species may die before they can reproduce, be sterile, or their offspring may be inviable.

REPRODUCTIVE BARRIERS

DYING OUT

The fate of a population depends on the balance between births and deaths: if birth rate exceeds death rate, populations swell; if deaths occur more often, populations shrink and may disappear altogether. Genes play their role, as some mutations cut short survival or fertility. Some harmful alleles are recessive (see pp.26–27) – two copies must come together for their harmful effects to be expressed. This is more likely to happen in populations that are already small because crosses between family members (who share similar genes) are more probable. This so-called inbreeding depression can increase the risk of population collapse in species already threatened by extinction.

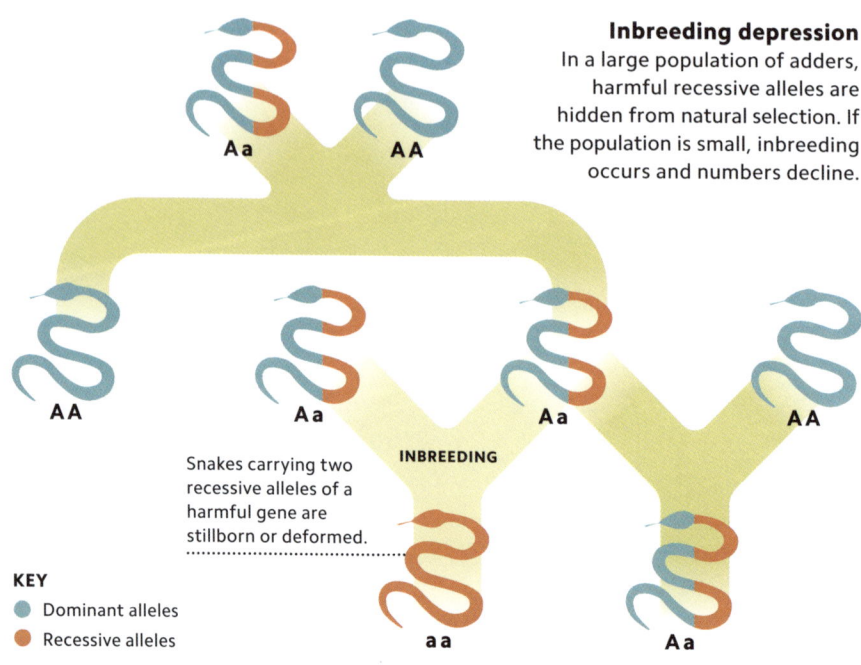

Inbreeding depression
In a large population of adders, harmful recessive alleles are hidden from natural selection. If the population is small, inbreeding occurs and numbers decline.

Snakes carrying two recessive alleles of a harmful gene are stillborn or deformed.

KEY
- Dominant alleles
- Recessive alleles

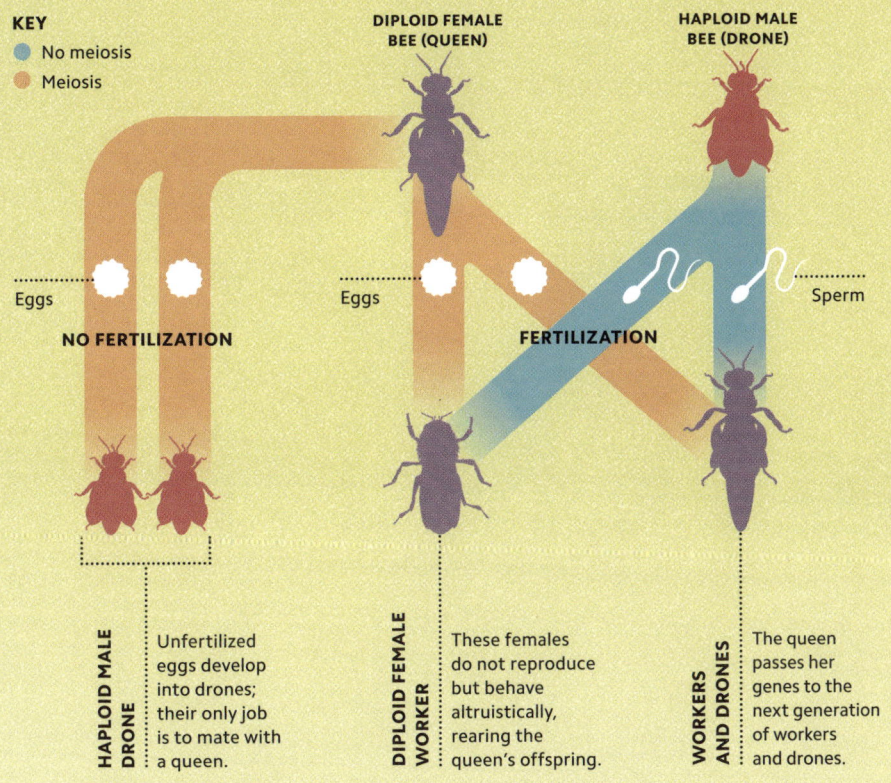

SELFISH GENES

Natural selection works by screening individuals according to their levels of fitness (see pp.122–23). One interpretation of this evolutionary process sees genes, rather than individuals, as the ultimate targets of selection; each gene has a "selfish" imperative to replicate itself and be passed on to the next generation. An argument in support of this idea comes from observations of altruism in colonial species. Female worker bees in a colony cannot breed, but they toil for the benefit of the colony. According to the selfish gene theory, they do so because they share genes with the queen, so their genes get passed on through her.

MANIPU
GENES

Genetic science is being harnessed to solve real human problems. Studies of genes and the proteins they produce help to diagnose and treat disease, solve crimes, and reveal family relations. Genetic engineering has become a common tool in biotechnology. Scientists are able to transfer genes that code for favourable characteristics between organisms, enabling bacteria to make medicines and crops to resist disease. Many other techniques are under development. Gene therapy aims to deliver healthy genes into a patient's cells to replace faulty ones, offering the potential to cure genetic disease; and through the cloning of DNA, extinct organisms may even be brought back to life.

PROTEIN SEQUENCING

Determining the sequence of DNA that encodes a gene is a difficult task, not least because DNA has a complex structure and is present only in tiny quantities in a cell. Proteins – the biomolecules formed by gene action – are more plentiful, easier to isolate, and have a simpler structure. Knowing the amino acid sequence of a protein allows the sequence of the DNA to be worked out (see p.89). Techniques for determining amino acid sequences involve splitting amino acids from the protein chain one by one and identifying them in turn.

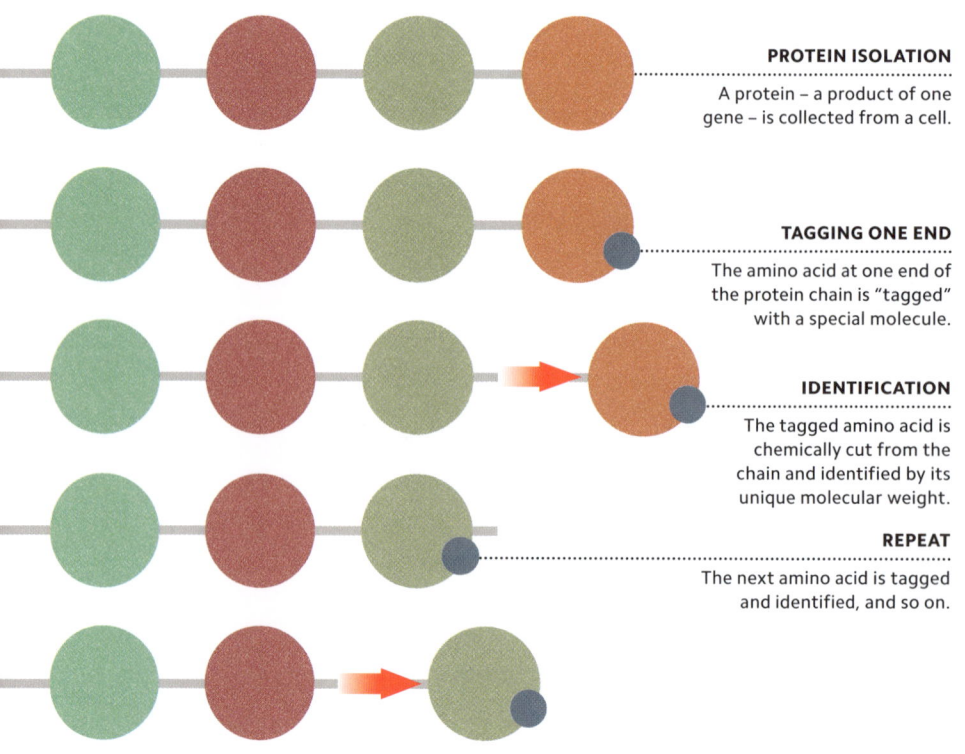

PROTEIN ISOLATION

A protein – a product of one gene – is collected from a cell.

TAGGING ONE END

The amino acid at one end of the protein chain is "tagged" with a special molecule.

IDENTIFICATION

The tagged amino acid is chemically cut from the chain and identified by its unique molecular weight.

REPEAT

The next amino acid is tagged and identified, and so on.

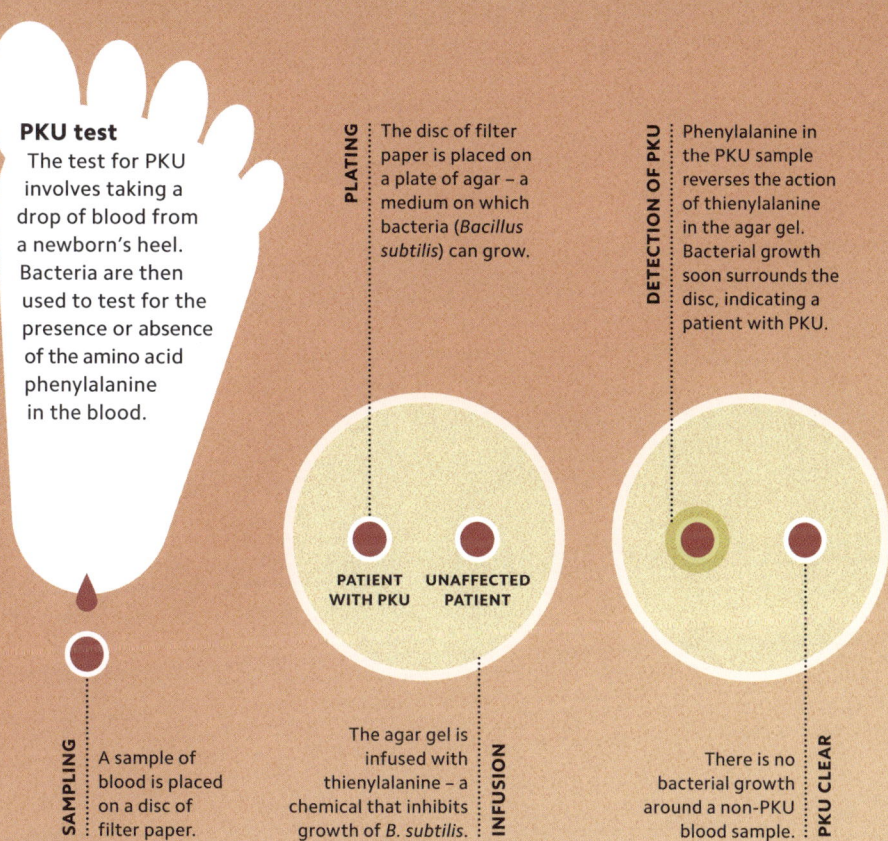

PKU test
The test for PKU involves taking a drop of blood from a newborn's heel. Bacteria are then used to test for the presence or absence of the amino acid phenylalanine in the blood.

PLATING — The disc of filter paper is placed on a plate of agar – a medium on which bacteria (*Bacillus subtilis*) can grow.

DETECTION OF PKU — Phenylalanine in the PKU sample reverses the action of thienylalanine in the agar gel. Bacterial growth soon surrounds the disc, indicating a patient with PKU.

SAMPLING — A sample of blood is placed on a disc of filter paper.

INFUSION — The agar gel is infused with thienylalanine – a chemical that inhibits growth of *B. subtilis*.

PKU CLEAR — There is no bacterial growth around a non-PKU blood sample.

FINDING BAD GENES

Some tests for genetic diseases in newborns depend on finding abnormal proteins or amino acids rather than directly probing the DNA sequence of the individual. These "marker proteins" point to the presence of genes that code for the genetic disease. One such test is for a metabolic disease called phenylketonuria (PKU). Sufferers of this disease cannot break down the amino acid phenylalanine because they have a mutation that stops them producing the enzyme phenylalanine hydroxylase. Phenylalanine builds up in the blood and brain, sometimes leading to brain damage.

MEDICAL GENE TESTS

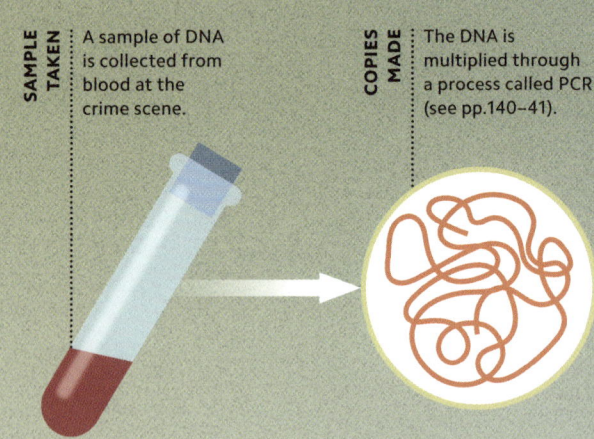

SAMPLE TAKEN — A sample of DNA is collected from blood at the crime scene.

COPIES MADE — The DNA is multiplied through a process called PCR (see pp.140–41).

Matching DNA
Genetic fingerprinting can be used to confirm identity and genetic relatedness. Here, DNA left behind at a crime scene (in blood, for example) is matched with that of a suspect.

> The first criminal conviction that relied on DNA evidence was made in 1988.

MAKING A MATCH

DNA analysis initially focused on identifying similarities between DNA samples. This process, called DNA profiling or "fingerprinting", relies on enzymes to cut DNA at particular base sequences, producing many fragments, which are then separated by gel electrophoresis. The fragments, which carry a slight electrical charge, are placed on a gel. A current is run through the gel, causing the fragments to move through it. The distance they move depends on their size, resulting in a series of bands in the gel. Comparison of these bands allows comparison of samples of DNA from individuals.

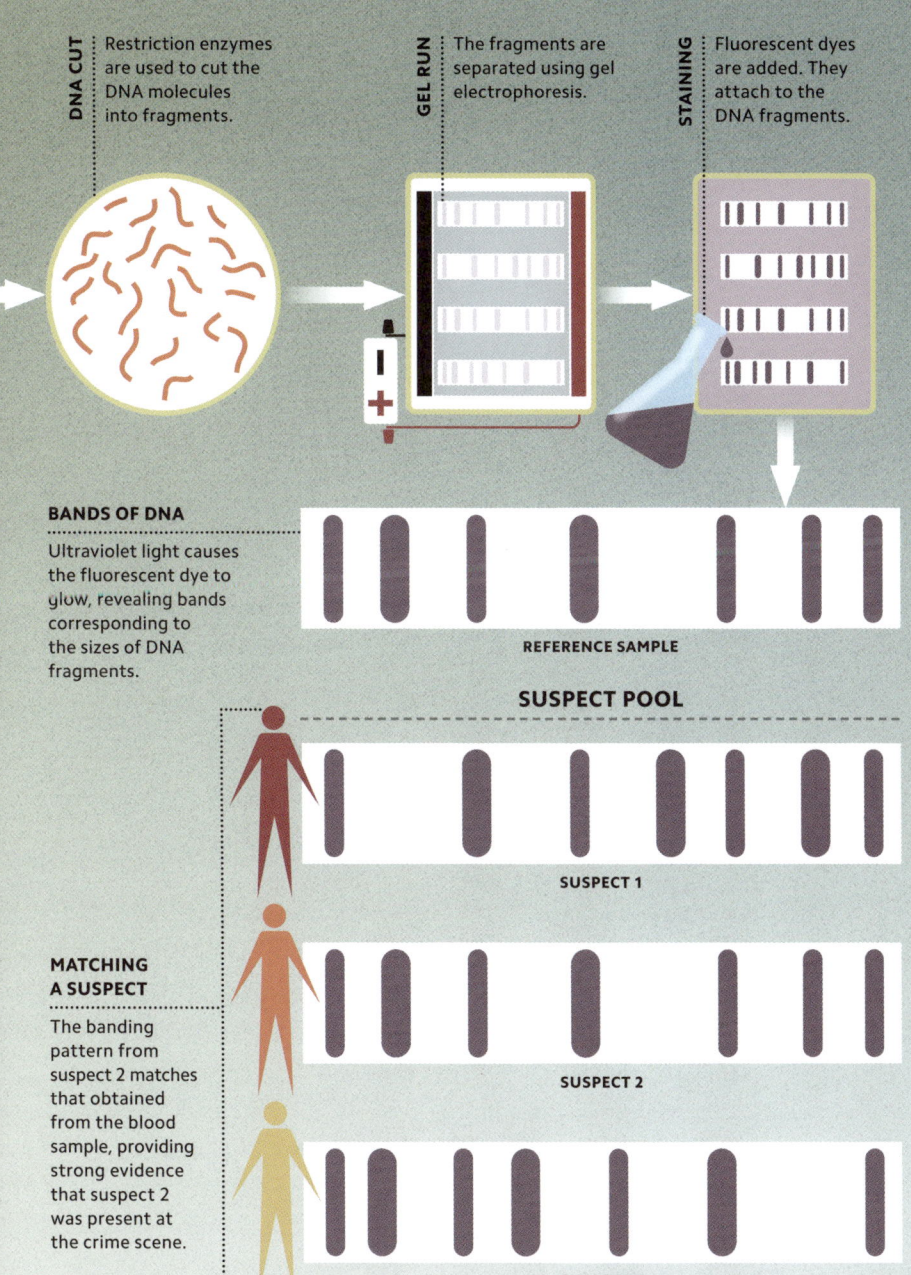

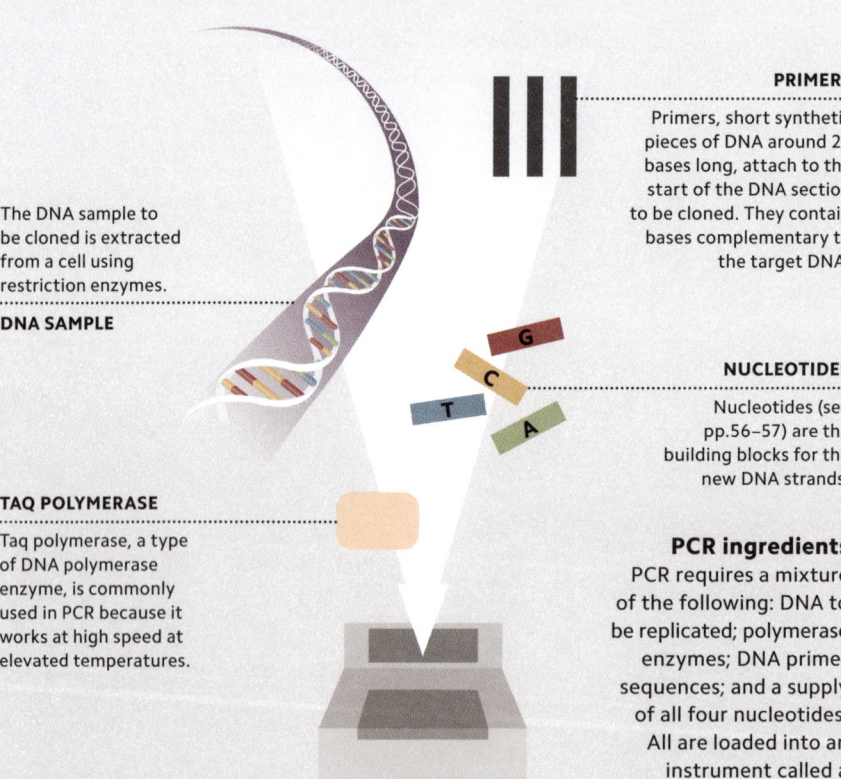

DNA SAMPLE

The DNA sample to be cloned is extracted from a cell using restriction enzymes.

PRIMERS

Primers, short synthetic pieces of DNA around 20 bases long, attach to the start of the DNA section to be cloned. They contain bases complementary to the target DNA.

NUCLEOTIDES

Nucleotides (see pp.56–57) are the building blocks for the new DNA strands.

TAQ POLYMERASE

Taq polymerase, a type of DNA polymerase enzyme, is commonly used in PCR because it works at high speed at elevated temperatures.

PCR ingredients

PCR requires a mixture of the following: DNA to be replicated; polymerase enzymes; DNA primer sequences; and a supply of all four nucleotides. All are loaded into an instrument called a thermal cycler.

THERMAL CYCLER

CLONING DNA

DNA is present in cells in tiny amounts. Amplifying – making multiple copies of the DNA molecule – is essential to gene analysis and engineering. An amplification technique, called the polymerase chain reaction (PCR), was developed in the 1980s. It allows scientists to clone isolated DNA in the laboratory by adding nucleotide building blocks and polymerases (DNA-making enzymes) to the DNA sample. The system uses a cycle of temperature changes to initiate a chain reaction in which copies of DNA molecules double in number with each round.

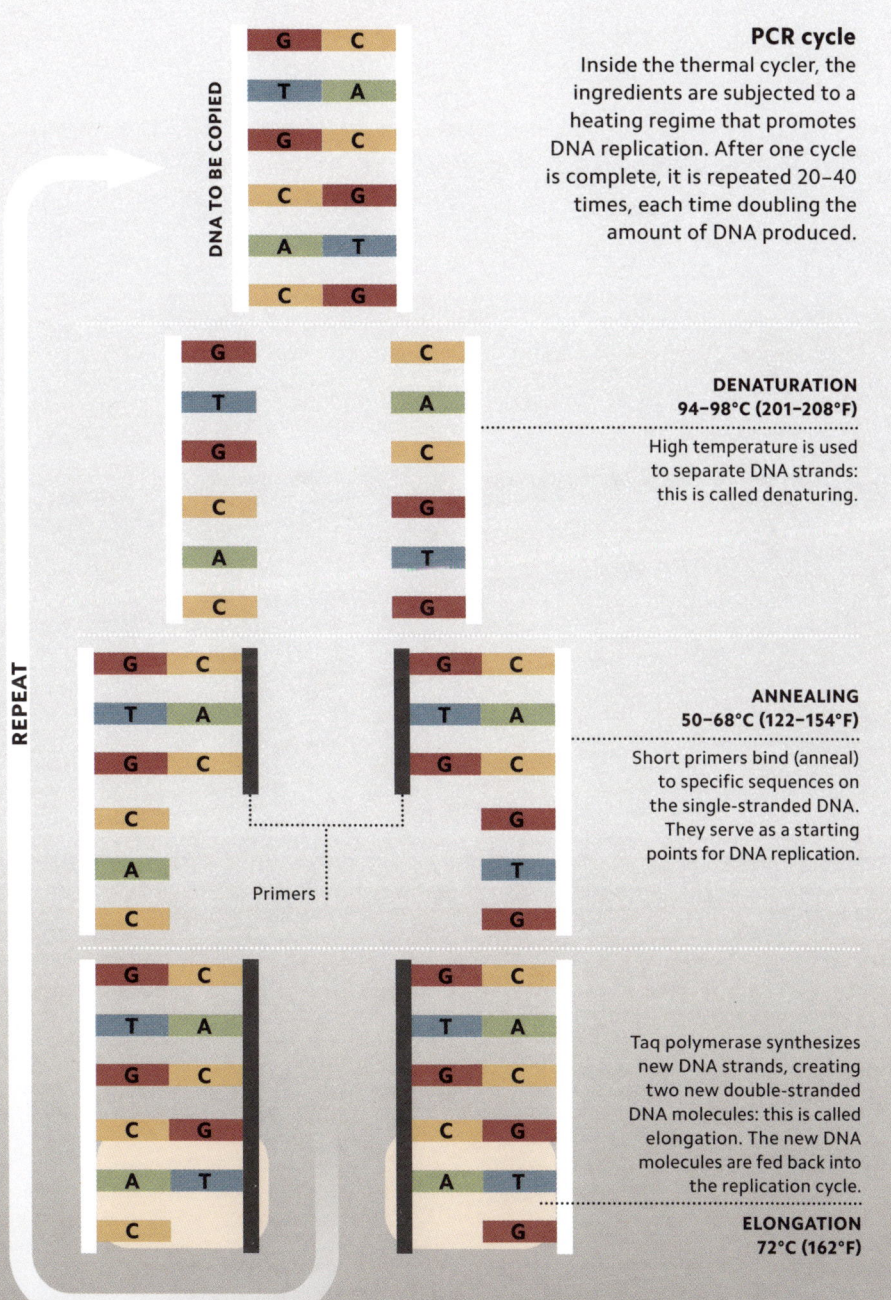

BASE TO BASE

It is much harder to sequence DNA directly than it is to sequence a protein. The molecules are fragile, and it is a challenge to cleave individual bases from a DNA molecule one by one (the technique that allowed proteins to be sequenced; see p.136). An innovative method developed in the 1970s by Frederick Sanger got around this problem by instead studying the sequence of bases added as DNA was replicated. The technique is known as Sanger sequencing, and it resulted in the first-ever complete sequence of an organism – that of a virus.

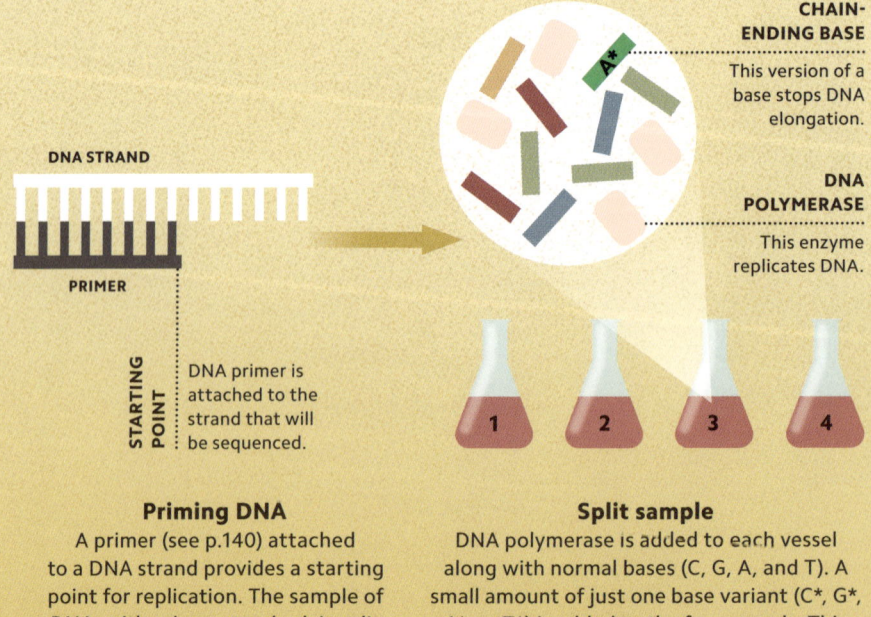

CHAIN-ENDING BASE
This version of a base stops DNA elongation.

DNA POLYMERASE
This enzyme replicates DNA.

DNA primer is attached to the strand that will be sequenced.

Priming DNA
A primer (see p.140) attached to a DNA strand provides a starting point for replication. The sample of DNA, with primer attached, is split into four; each sample is placed into a separate vessel.

Split sample
DNA polymerase is added to each vessel along with normal bases (C, G, A, and T). A small amount of just one base variant (C*, G*, A*, or T*) is added to the four vessels. This variant stops the polymerization reaction as soon as it is incorporated into a DNA chain.

> DNA sequencing is central to many medical applications, including cancer diagnosis and the development of new antibiotic drugs.

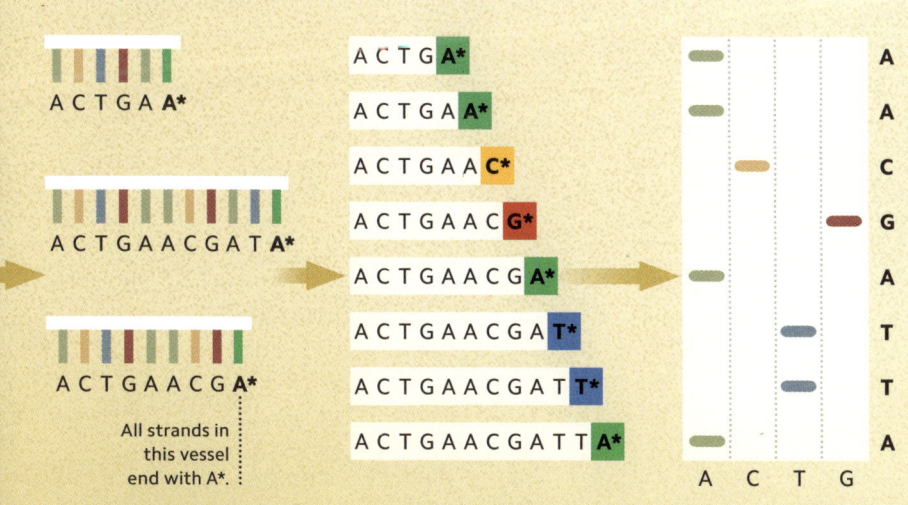

Ending the chain
The chain-ending base (in this case A*) is occasionally incorporated into the strand, ending elongation. The result is a vessel full of DNA strands of various lengths.

C, G, A, T fragments
The process is repeated in the other three vessels, using T*, C*, or G* as the chain-ending base. The strands are separated by electrophoresis (see p.138).

Reading the DNA
Knowing the size of each DNA fragment, and the base with which it is terminated, allows the sequence of the DNA to be determined.

DNA SEQUENCING | 143

MAPPING DIVERSITY

Automated machines are capable of carrying out DNA sequencing at ever-increasing speeds, allowing the largest genomes to be analysed. Such instruments were first used to sequence the genomes of microbes, simple worms, and fruit flies. The first draft of the human genome was published in 2003, and the full version in 2022. This monumental achievement identified the positions of more than 20,000 genes on 23 pairs of human chromosomes across a sequence of 3.2 billion bases.

BACTERIOPHAGE MS2
The first whole genome to be mapped (in 1976) was of a virus that infects bacteria; it has only four genes.

HAEMOPHILUS INFLUENZAE
The first free-living organism to have its genome mapped was this bacterium; it has over 1,700 genes.

SACCHAROMYCES CEREVISIAE
The first eukaryote to have its genome sequenced was yeast, in 1996. It has more than 6,000 genes.

CAENORHABDITIS ELEGANS
This nematode worm was the first animal to have its genome sequenced. It has around 20,000 genes.

Sequences revealed
The organisms targeted for genome sequencing include those traditionally used in genetic studies.

DROSOPHILA MELANOGASTER
The fruit fly genome was published in 2000; it has some 13,600 genes.

HOMO SAPIENS
The Human Genome Project has led to many advances in medicine.

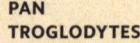

PAN TROGLODYTES
The chimpanzee genome was sequenced by 2013, revealing great similarities with humans.

PUTTING GENETICS TO WORK

Humans have been improving crops and domesticated animals for millennia by selecting organisms with desirable characteristics and breeding them to amplify these traits. Selective breeding of wild cabbage, for example, has allowed growers to produce cauliflower, broccoli, and kohlrabi from a common ancestor. More recently, an understanding of DNA, genes, and the enzymes controlling them has opened up genetic engineering techniques for cutting individual genes from cells and moving them, even between different species.

Traditional method
Selective breeding techniques are – genetically – a blunt instrument. They may move a desired gene into a recipient organism, but along with many other, unwanted, genes.

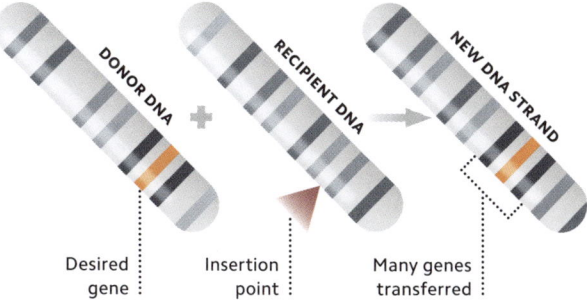

Desired gene · Insertion point · Many genes transferred

Genetic engineering
New techniques allow one or a few genes to be inserted into the target organism with precision. Genetically modified organisms (GMOs) include cotton plants carrying a gene that produces a toxin to kill parasitic bollworms.

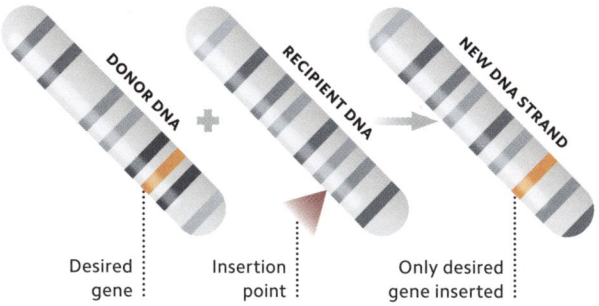

Desired gene · Insertion point · Only desired gene inserted

MICROBE MECHANICS

Genetic engineering opened up many possibilities for changing the genomes of bacteria, plants, and animals. One of its first applications was to alter the genes in a bacterium to induce it to make human insulin – a hormone needed to control diabetes. Previously, insulin had to be extracted from the pancreas of a pig or cow. The transformation was achieved by splicing an insulin gene into a bacterial plasmid – a small circular DNA molecule that is separate from the bacterial chromosome.

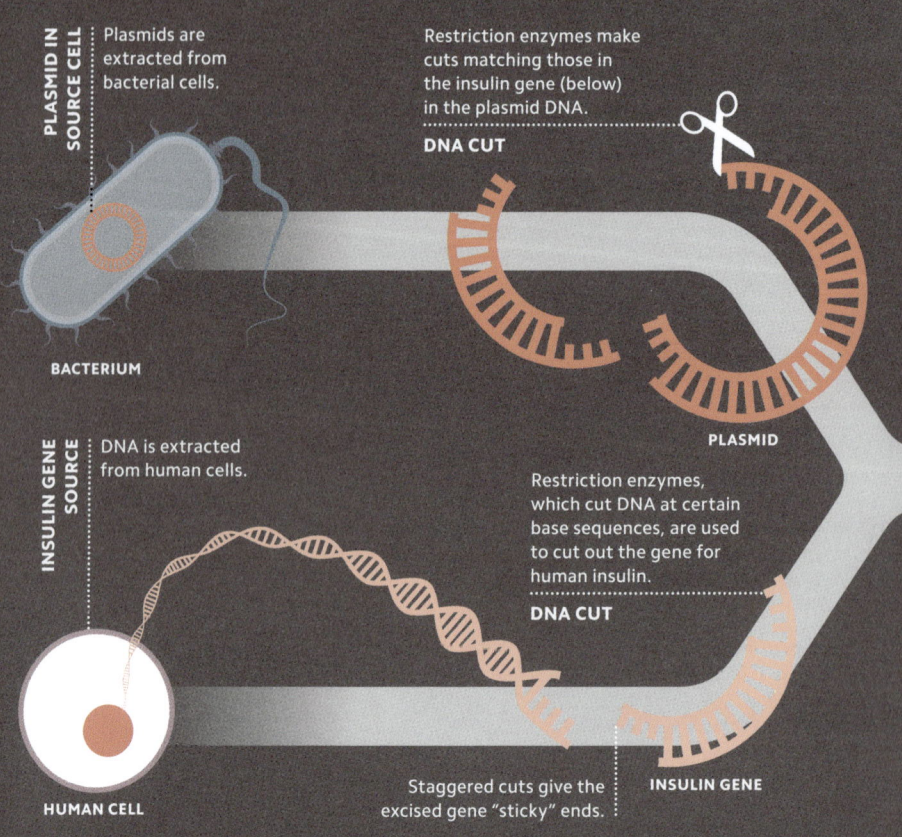

PLASMID IN SOURCE CELL — Plasmids are extracted from bacterial cells.

BACTERIUM

DNA CUT — Restriction enzymes make cuts matching those in the insulin gene (below) in the plasmid DNA.

PLASMID

INSULIN GENE SOURCE — DNA is extracted from human cells.

HUMAN CELL

DNA CUT — Restriction enzymes, which cut DNA at certain base sequences, are used to cut out the gene for human insulin.

INSULIN GENE

Staggered cuts give the excised gene "sticky" ends.

146 | ENGINEERING BACTERIA

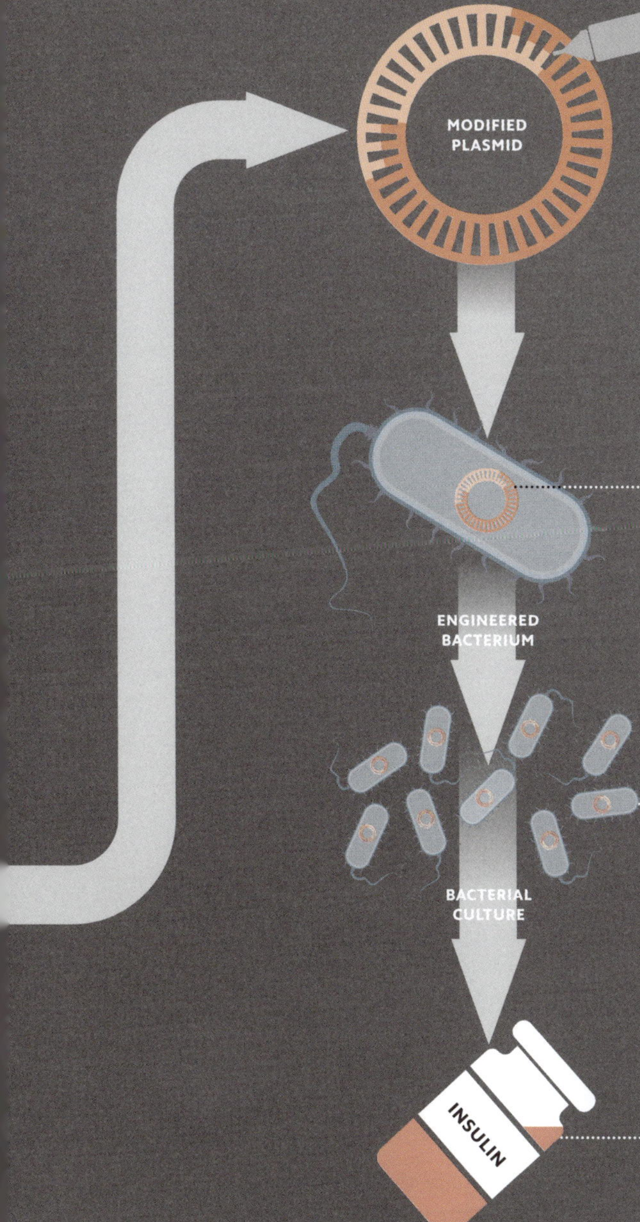

GENE SPLICING
Gluing enzymes called ligases recognize the staggered, complementary cuts in the DNA and seal the insulin gene within the plasmid.

MODIFIED PLASMID

MODIFIED GENE INSERTED
Bacterial cells are "shocked" by high temperature to induce them to take up the genetically modified plasmids that carry the "foreign" insulin gene.

ENGINEERED BACTERIUM

INSULIN FACTORIES
The bacteria are cultured in great numbers in the laboratory.

BACTERIAL CULTURE

PRODUCT HARVESTED
The insulin harvested is the same as that made in the human body by the pancreas.

ENGINEERING BACTERIA | 147

CARRIERS OF CHANGE

Plasmids are small circular rings of DNA that occur naturally in bacteria, archaea, and some fungi. They are used in genetic engineering to carry genes from one organism to another. They can move genes into bacterial cells (see pp.146–47) as well as into the more complex cells of plants and animals. However, plasmids are not the only gene vectors used by geneticists. Viruses and oily spheres called liposomes are also put to work in engineering plants and animals for desirable properties, such as boosting their yield or improving resistance to disease.

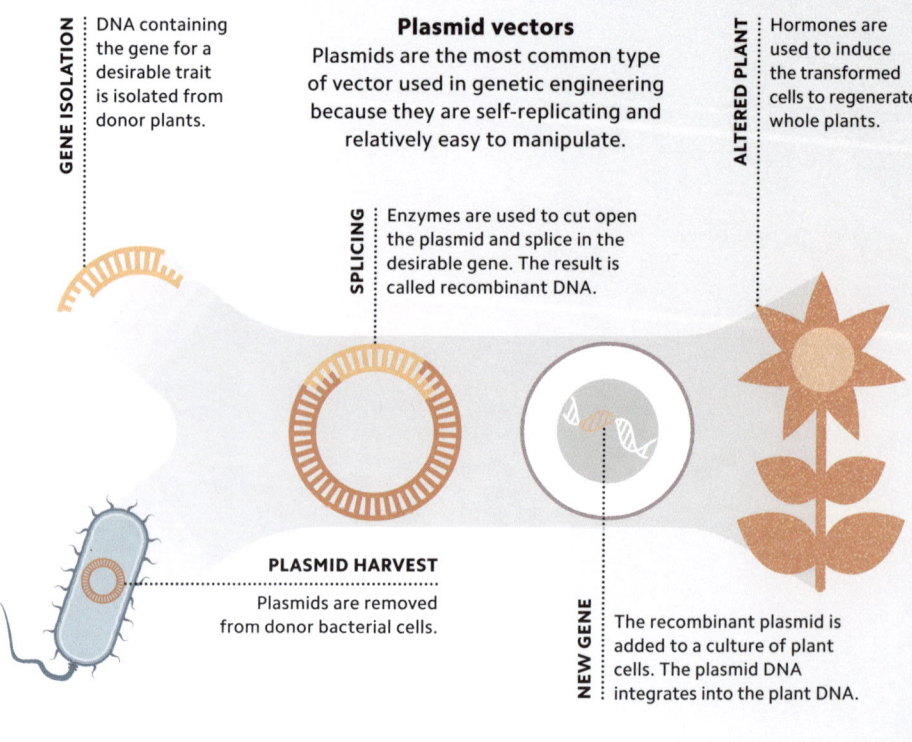

GENE ISOLATION
DNA containing the gene for a desirable trait is isolated from donor plants.

Plasmid vectors
Plasmids are the most common type of vector used in genetic engineering because they are self-replicating and relatively easy to manipulate.

ALTERED PLANT
Hormones are used to induce the transformed cells to regenerate whole plants.

SPLICING
Enzymes are used to cut open the plasmid and splice in the desirable gene. The result is called recombinant DNA.

PLASMID HARVEST
Plasmids are removed from donor bacterial cells.

NEW GENE
The recombinant plasmid is added to a culture of plant cells. The plasmid DNA integrates into the plant DNA.

148 | GENE VECTORS

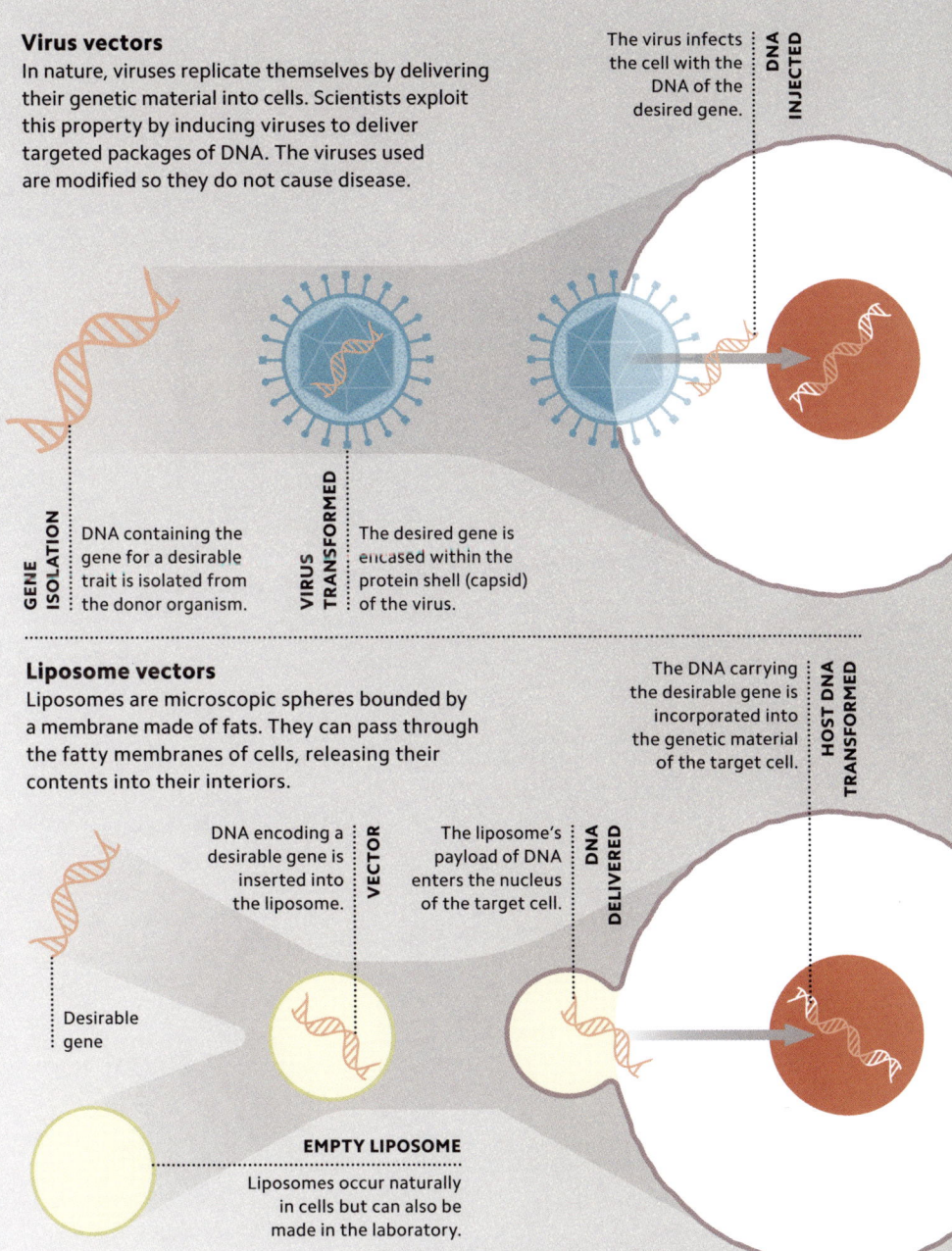

PASSING IT ON

Genetic modifications made to germ cells (sex cells) or early-stage embryos are inherited by all the cells in an organism's body and are subsequently passed on to its offspring. This type of genetic editing is known as germline modification; it is highly restricted in human cells in most countries on ethical grounds. The genetic engineering of non-germline, or somatic, cells, is the basis of gene therapy (see opposite). Somatic modifications cannot be passed on to offspring, and affect only limited cell types.

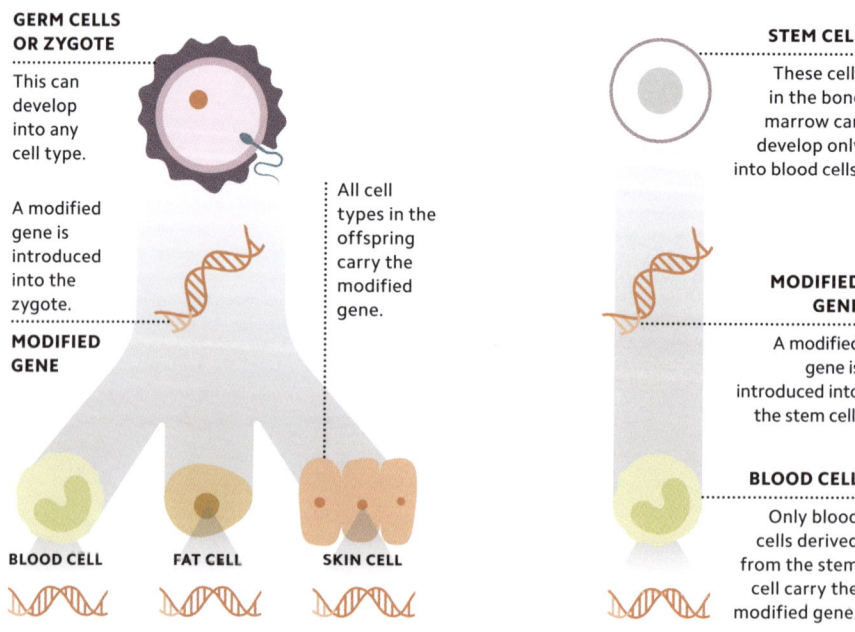

GERM CELLS OR ZYGOTE
This can develop into any cell type.

A modified gene is introduced into the zygote.

MODIFIED GENE

All cell types in the offspring carry the modified gene.

BLOOD CELL **FAT CELL** **SKIN CELL**

STEM CELL
These cells in the bone marrow can develop only into blood cells.

MODIFIED GENE
A modified gene is introduced into the stem cell.

BLOOD CELL
Only blood cells derived from the stem cell carry the modified gene.

Germline modification
Here a gene is spliced into the zygote (a fertilized egg). It is passed on to every cell, and every cell type, in the body.

Somatic modification
Here, a gene is spliced into a stem cell in the bone marrow. Stem cells can develop into only a limited range of cell types.

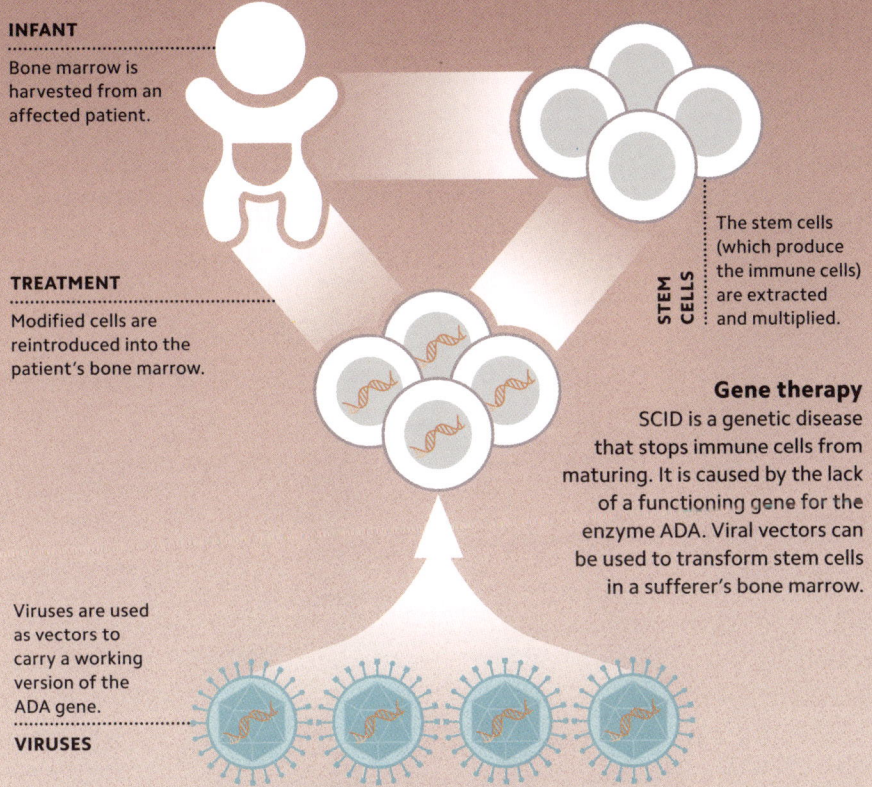

INFANT
Bone marrow is harvested from an affected patient.

TREATMENT
Modified cells are reintroduced into the patient's bone marrow.

Viruses are used as vectors to carry a working version of the ADA gene.

VIRUSES

STEM CELLS
The stem cells (which produce the immune cells) are extracted and multiplied.

Gene therapy
SCID is a genetic disease that stops immune cells from maturing. It is caused by the lack of a functioning gene for the enzyme ADA. Viral vectors can be used to transform stem cells in a sufferer's bone marrow.

HEALING GENES

In gene therapy, a functional gene is inserted into cells of parts of the body affected by a genetic disease. The target could be an organ such as the lungs, to treat cystic fibrosis, or stem cells like those in bone marrow to treat disorders such as severe combined immunodeficiency (SCID). An alternative strategy, called antisense therapy, aims to treat diseases by blocking a cell's ability to make RNA that codes for a certain protein, so preventing harmful genes, such as cancer-inducing genes, from being active.

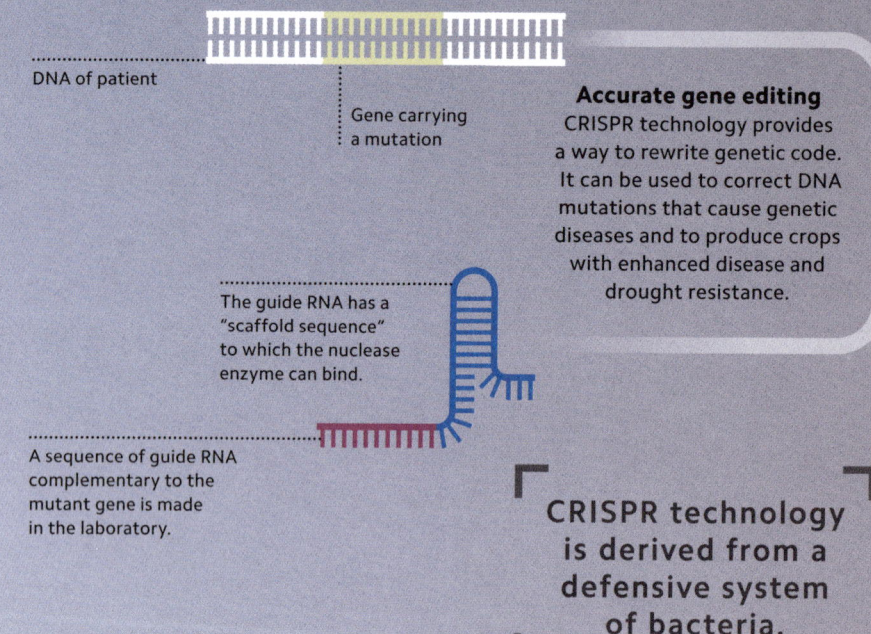

DNA of patient

Gene carrying a mutation

Accurate gene editing
CRISPR technology provides a way to rewrite genetic code. It can be used to correct DNA mutations that cause genetic diseases and to produce crops with enhanced disease and drought resistance.

The guide RNA has a "scaffold sequence" to which the nuclease enzyme can bind.

A sequence of guide RNA complementary to the mutant gene is made in the laboratory.

> CRISPR technology is derived from a defensive system of bacteria.

REWRITING GENES

CRISPR is a sophisticated gene-editing technology capable of disabling harmful genes and even repairing faulty ones. "CRISPR" refers to DNA sequences in bacteria that help them defend against viral attack. CRISPR sequences are transcribed into RNA (see pp.90–91); in the bacterial cell, this RNA guides enzymes called nucleases to cut certain sequences of the viral DNA, so defeating the attacker. In 2012, geneticists found a way to manufacture CRISPR RNA molecules that could guide nucleases to almost any desired DNA sequence. This provided a powerful tool for targeting and cutting DNA with great precision.

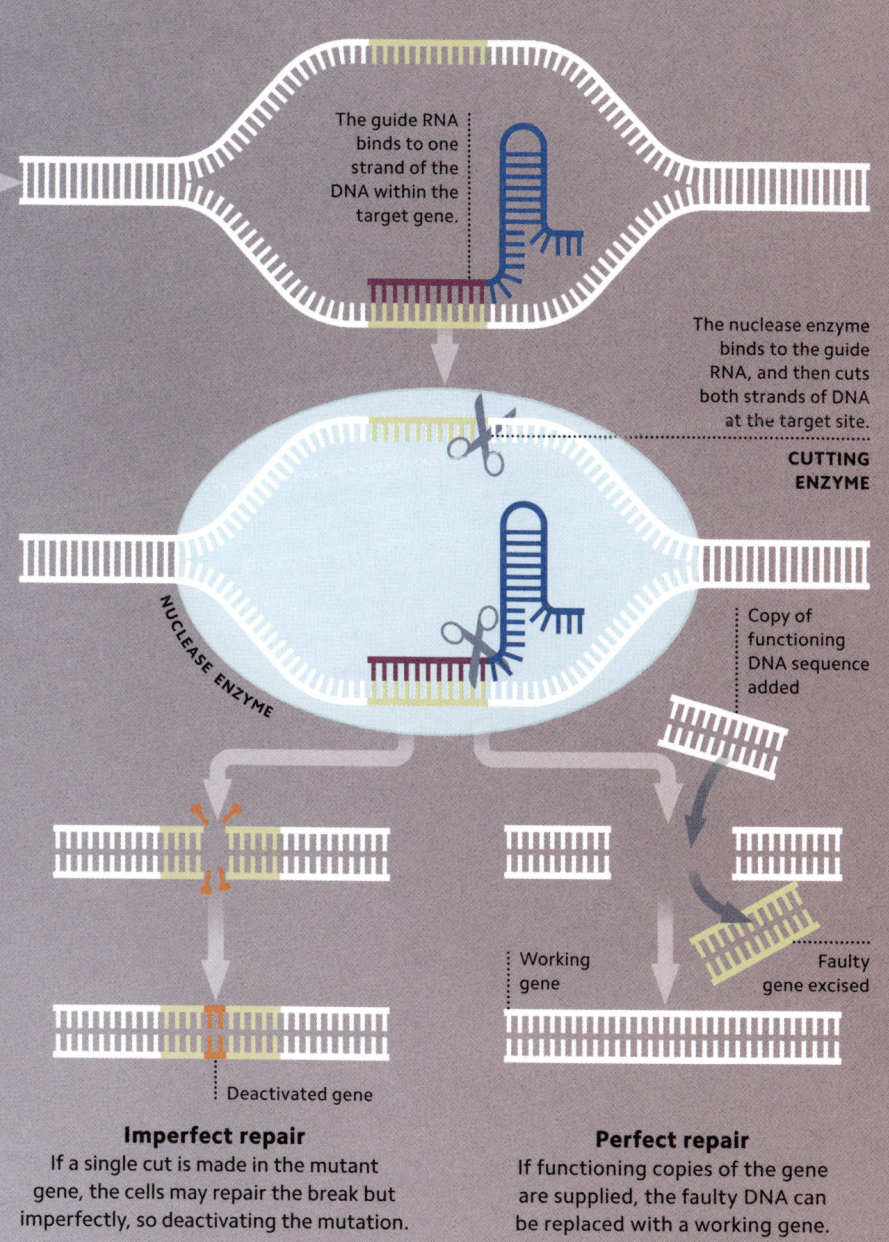

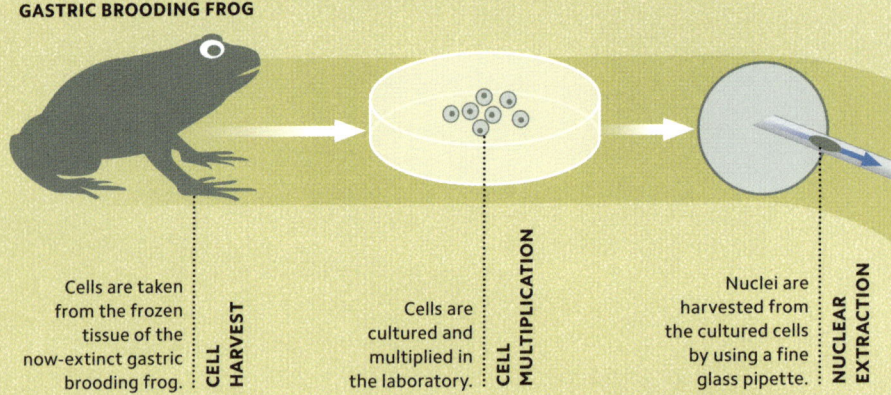

DE-EXTINCTION

Advances in biotechnology have given scientists the tools to clone living things, producing identical copies of a single parent from somatic cells (see p.150). The first mammal to be cloned using a technique called somatic cell nuclear transfer (SCNT) was Dolly, a sheep, in 1996; many more species – from arctic wolves to macaque monkeys – have followed. SCNT can potentially be used to resurrect extinct species from their preserved remains, although attempts to do so have so far fallen short of complete success.

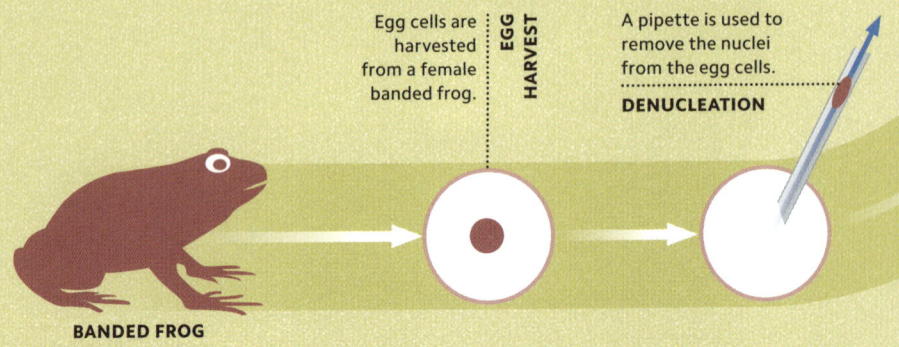

Somatic cell nuclear transfer (SCNT)
The Australian gastric brooding frog became extinct in the 1990s. Scientists set out to resurrect a frog from frozen tissue by inserting its DNA into an egg cell of a related frog species, from which the nucleus had been removed.

EMBRYO FORMATION
An embryo forms from the manipulated cell.

NUCLEAR INJECTION
The harvested nuclei are injected into the denucleated egg cells.

DEVELOPMENT
A tadpole develops into a gastric brooding frog.

ADULT FROG
The gastric brooding frog matures into an adult.

> In 2017, the cynomolgus (crab-eating) monkey became the first primate to be successfully cloned.

CLONING AND RESURRECTION

INDEX

Page numbers in **bold** refer to main entries.

A

activator proteins 99
adenine (A) 58–59, 60, 95, 142–43
albinism 107
alleles **25–27**, 116, 117, 123
  blood types 49
  characteristics 50, 52
  crossing over 42, 45, 78
  mechanisms of evolution 118, 119, 121
  mutations 112, 113
allopatric speciation **128–29**
amino acids 63, 85, 86, 88
  abnormal 137
  determining sequences 136
  mutations 106, 110
  translation **92–93**
  triplet code **89**
amoeba 22
anaphase 69, 79, 80
annealing 141
antibodies 17, 85
antisense therapy 151
archaea 12, 148
asexual reproduction 18, **22**, 74

B

bacteria 14, 74, 96, 98, 104, 144
  asexual reproduction 22
  CRISPR **152–53**
  engineering bacteria **146–47**
  plasmids 146–47, 148
bacteriophage MS2 144
barriers, reproductive **131**
bases 56, 57, 63
  base pairs 58, 59, 60, 61, 70

base sequences 59, 61, 142–43, 144
  Human Genome Project **144**
  mutations 106, **107**, 110–11
  translation 92, 93
bees 18, 133
beta haemoglobin 110–11
birds 126, 131
  blood types 49
bone marrow 150, 151
bottlenecks, genetic 118, **121**
bowerbirds 126
bread mould 84
breeding **11**, 145
  experiments 32–41, 44–45
  interbreeding 11, 116, 117, 131, 132

C

Caenorhabditis elegans 144
cancer 104, 143, 151
carbon 56, 57, 86
catalysts **87**
cats 26, 53
cattle 27
cells 10, **14–15**, 22, 66, 76, 87
  cell cycle **66**
  cell division 66, **67**, **68–69**, 70, 76, **78–81**, 109
  central dogma **88**
  drivers of early development **101**
  genes and cell activity **17**
  interphase 66, 68
  meiosis **67**, **78–79**
  mitosis 66, **67**, **68–69**
  nucleus 62, 65, 68, 78
  somatic cells 150, 154
  see also sex cells

centromeres 64, 68, 69, 80
centrosomes 69, 79, 80
characteristics 48, 49, **50**, 51, 74, 145
  selection and **122–26**
chemicals 104
chimpanzees 21, 144
chromatids 68, 69, 80, 81
chromatin 62
chromosomes **16**, 64, 65, 90
  chromosome theory of inheritance **40–41**
  crossing over 42, 45
  diploidy **76–77**
  genetic diseases **112–13**
  haploidy **76–77**
  homologous pairs 24, 42, 44, 78–79, 109
  Human Genome Project **144**
  linkage maps **43**
  meiosis 67, 78, 108
  mitosis 66, 67, **68–69**
  mutations 104, **108–109**, 130
  non-coding DNA 25
  number of **18**, 62, 70, 71
  patterns **20–21**
  polyploidy 130
  and sex cells **24**
  sex chromosomes 16, **19**
species 16, 18, 20
cloning **140–41**, **154–55**
codons 89, 92, 107
collagen 85
crime scenes 138–39
CRISPR **152–53**
crops 145, 152
crosses, dihybrid **38–39**
crossing over 42, 45, 78
cystic fibrosis 112, 113, 151
cytokinesis 66, 69, 79, 81

cytoplasm 66, 90, 91
cytosine (C) 58–59, 60,
    142–43

D

Darwin, Charles 122
deletion 109
denaturation 141
deoxyribose 56, 57
differentiation 101
diffusion 101
dihybrid crosses **38–39**
diploid **76–77**, 78, 81, 130
directional selection 124, **125**
diseases **112–13**, 121, 137,
    148, **151–53**
disruptive selection 124, **125**
diversity 23, **74–75**, 144
DNA (deoxyribonucleic acid)
    7, 14, 17, 54–71, 88
  cell cycle 66
  cloning **140–41**, **154–55**
  coiling and packaging 62
  CRISPR **152–53**
  definition of 16
  DNA fingerprinting
    138–39
  double helix 16, **58–59**, 60,
    62, 90
  epigenetic factors 100
  frameshift mutations **107**
  molecules of inheritance
    56–57
  mutations 104, 106, 119
  non-coding **64**, 94
  sequencing **142–43**
  templates and replication
    60–61, 66, 68, **70–71**, 78,
    88, 90, 106, 120
  see also bases;
    chromosomes, mRNA;
    RNA
DNA polymerase 71, 140, 142
Dolly, the sheep 154
dominant genes 37, 38, 39,
    40, 48, 50, 52
donkeys 11
Down's syndrome 108
duplication 109

EF

eggs 23, 30, 31, 67, 76, 78,
    81, 101
electrophoresis, gel 138, 139,
    143
elongation 141, 142, 143
embryos 23, 101, 108, 150
endoplasmic reticulum 91
engineering organisms **145**
environmental factors 51, **53**
enzymes 17, 71, 84, 85, 96
  DNA fingerprinting 138
  plasmids 148
  "proofreading" 106
  and proteins **87**
  restriction enzymes 146
epigenetics **100**
epistasis **52**
equilibrium, genetic **117**, 119
eubacteria 12
eukaryotic cells 15, 16, 17,
    65, 98, 144
evolution 12, **114–33**
exons 64, 94–95
extinction 13, **132**
  de-extinction **154–55**
eye colour 52, 84
factors 45
  epigenetic factors **100**
  inheritable **31**, 32, **36–41**,
    42
fertilization 23, 24, 75, 76, 77
fingerprinting, DNA **138–39**
founder effect **121**
four o'clock plant 32–33
frameshift mutations **107**
frogs 131, 154–55
fruit flies **40–41**, 42, 43, 144
fungi 12, 14, 76, 148

G

gametes see sex cells
gene pools 116
genes 7, 16, 64, 90
  alleles **25–27**
  analysing genes via
    proteins **136**
  and cell activity **17**

characteristics 48, 49, 50,
    51
  and chromosomes **16**
  in combination **50–51**
  dominant vs recessive
    26–27
  drivers of early
    development **101**
  epigenetics **100**
  epistasis **52**
  gene regulation **96–97**
  gene therapy 150, **151**
  gene vectors **148–49**
  how genes work **82–101**
  and life **10**
  linked genes **42–43**
  manipulating **134–55**
  medical gene tests **137**
  and reproduction 24,
    72–81
  switching on and off
    98–99, 100
  and variety **12–13**
  see also mutations
genetic code 59, 60
genetic drift 118
genetic engineering **134–55**
  cloning and resurrecting
    DNA **154–55**
  CRISPR **152–53**
  engineering bacteria
    146–47
  somatic and germline
    modification **150**
genetically modified
    organisms (GMOs) 145
Genome Project, Human
    144
genotypes **48**, 50, 116
germline modification **150**
germline mutation **105**
guanine (G) 58–59, 60,
    142–43

H

haemoglobin 110–11
haemophilia 112
Haemophilus influenzae 144
haploidy **76–77**, 78, 81

INDEX | 157

Hardy-Weinberg principle
117
hedgehog mutation 107
height 51
helicase 71
heterozygous 25
histones 62, 100
Homo sapiens 144
homologous chromosomes
24, 42, 44, 78–79, 109
homozygous 25
"hopeful monsters" 127
hormones 85, 99, 148
horses 11, 27
Human Genome Project
144
Huntington's disease 112, 113
hydrogen 56, 58, 86

I

infections 104
inheritance **28–45**, 60, 117
3:1 inheritance **36–37**
chromosome theory of
inheritance **40–41**
DNA as molecules of
inheritance **56–57**
genetic test beds 34–35
linked genes **42–43**
material inheritance **31**
mechanics of inheritance
32–33
physical basis of
inheritance **44–45**
preformation 30
ratios of inheritance
36–39
insulin 84, 85, 146–47
interbreeding 11, 116, 117,
131, 132
interphase 66, 68
introns 64, 94–95
inversion 109

JKL

Johannsen, Wilhelm 7
karyotypes **20–21**
katydids 122–23

kingdoms **12–13**
lac operon **96–97**
lactase 96, 97, 120
lactose 96, 97, 120
lagging strands 71
leading strands 71
life: diversity of **12–13**
genes and **10**
ligases 147
linkage maps 43
liposomes 148, **149**

M

material inheritance **31**
medical gene tests **137**
meiosis **67**, **76–81**, 108, 130
melanin 107
Mendel, Gregor 7, **34–39**,
40, 41, 42, 45
messenger RNA (mRNA) 63,
88, **90–93**, 95, 96
metaphase 69, 78, 80
microtubules 68, 69, 78, 79
mitochondria 65
mitochondrial DNA
(mtDNA) **65**
mitosis 66, **67–69**, 76–77
monkeys 155
Morgan, Thomas **40–41**, 42
moss 77
motor proteins 85
mules 11
mutagenic agents 104
mutations 64, 74, 75,
102–113, 119, 127
causes of **104**
evolution by **120**
extinctions and 132
frameshift mutations **107**
and gene function **110–11**
genetic diseases **112–13**
somatic and germline
mutation **105**
structural chromosome
mutation **109**
substitution mutation **106**
sympatric speciation **130**
whole chromosome
mutation **108**

NO

natural selection 119,
122–23, 124, 126, 132,
133
nitrogen 56, 86
non-coding DNA (ncDNA) 64
nuclear membranes 68, 69,
79, 91, 92
nuclease enzymes 152, 153
nucleotides 56–57, 61, 70,
71, 90, 140
nucleus 14, 15, 16, 56, 62,
65, 68, 78
operons 96, 98
organelles 66, 101
organic bases 56, 57, 58
organisms: engineering **145**
model **34–35**
oxygen (56, 57, 86

P

Pan troglodytes 144
peas 25, 34–39, 44–45
peptide bonds 86
phenotypes **48**, 50
phenylketonuria (PKU) 137
phosphates 56, 57, 58, 63
phosphorus (P) 56, 57
plants **12**, 13, 14, 77, 146
breeding barriers 131
polyploidy 130
reproduction 23, 74
plasmids 146, 147, **148**
polymerase chain reaction
(PCR) 138, **140–41**
polymers 56
polymorphism 50
polypeptides 86
polyploidy 130
population genetics **114–33**
preformation 30
primers 140, 141, 142
prokaryotes 14, 96, 98
promoter sequence 90
prophase 68, 78, 80
protein synthesis 60
proteins 17, 63, 66, 90, 98
abnormal 137

158 | INDEX

amino acids 86, 88, 89, 106, 136
analysing genes via **136**
and enzymes **87**
gene codes for 64, **84**
histone proteins 62, 100
introns 94
mitochondrial DNA (mtDNA) 65
mutations 106, **110–11**
protein production 60, 89, 91, **92–93**
protein sequencing **136**
structure of **86**
what proteins do **85**
protozoa 12, 14, 22

R

radiation 104
receptors 85
recessive genes 37, 39, 40, 48, 50, 52
recombinant DNA 148
redundancy 89
repetitive ncDNA 64
replication, DNA **60–61**, 66, 68, **70–71**, 78, 88, 106, 120
repressor proteins 96, 97
reproduction 10, **22–23**, **72–81**
  generating genetic diversity 23, **74–75**
  see also asexual reproduction; sexual reproduction
restriction enzymes 146
resurrecting DNA **154–55**
ribose 63
ribosomes 63, 88, 90, 91, 92–93
RNA (ribonucleic acid) 60, **63**, 151
  CRISPR **152–53**
  messenger RNA (mRNA) 63, 88, **90–93**, 95, 96
  RNA editing **94–95**
  transfer RNA (tRNA) 63, 92, 93

RNA polymerase 90–91, 96, 97, 100

S

Saccharomyces cerevisiae 144
Sanger, Frederick 142
selection **122–26**, 132, 133, 145
selfish gene theory **133**
sense strand 60
sequencing: DNA **142–43**
  proteins **136**
severe combined immunodeficiency (SCID) 151
sex cells (gametes) 22, 23, 30, 31, 75, 77, 126
  and chromosomes **24**
  diploid and haploid 76, 130
  formation of 42, 45, **80–81**
  meiosis 67
  mutations 105
  polyploidy 130
  somatic and germline modification **150**
sex chromosomes 16, **19**
sex-linked diseases **112**
sexual reproduction 22, 23, 24, 74, 75, 104
sexual selection **126**
sickle cell disease **110–11**, 112
snakes 132
somatic cells **105**, **150**, 154
somatic cell nuclear transfer (SCNT) **154–55**
species 12, 13, 119
  allopatric speciation **128–29**
  breeding **11**
  chromosomes 16, 18, 20
  cloning and resurrecting **154–55**
  micro- and macroevolution **127**
  reproductive barriers **131**
  sympatric speciation **130**
sperm 23, 30, 67, 76, 78, 81, 84

spindle fibres 69, 78, 80
spliceosomes 95
spores 77
sporophytes 77
stabilizing selection **125**
stem cells 150
sterility 11, 131
storage proteins 85
structural proteins 85
substitution mutation **106**
substrate 87
sugar molecules 56, 57, 58, 63
sympatric speciation **130**

T

tamarin species 128–29
taq polymerase 140
telomeres 64
telophase 69, 79, 81
templates **60–61**, 70
termination codons 89
tests **137**
thermal cyclers 140–41
thymine (T) 58–59, 60, 63, 142–43
totipotent cells 101
transcription 17, 64, **90–91**, 94, 96, 98
transduction 99
transfer RNA (tRNA) 63, 92, 93
translation 17, **92–93**
translocation 109
transport proteins 85
transposons 64
triplet code **89**, 110
triploid 108
trisomy 108
tRNA (tyrosinase 107

UVZ

uracyl (U) 63
variation **12–13**, **46–53**, 104
variegation 53
vectors, gene **148–49**
viruses 104, 142, 144, 148, **149**, 151
zygotes 23, 81, 150

INDEX | 159

ACKNOWLEDGMENTS

DK would like to thank the following for their help with this book: Christine Stroyan for her managerial read; Katie John for proofreading; and Vanessa Bird for the index.

Cover images: *Front* and *Back:* Shutterstock.com: Soleil Nordic

DK LONDON
Project Editors Marek Walisiewicz, Daniel Byrne
Project Art Editors Clare Joyce, Daksheeta Pattni
Designer Phil Gamble
Managing Editor Gareth Jones
Managing Art Editor Luke Griffin
Production Editor Robert Dunn
Production Controller Nancy-Jane Maun
Pre-Production Designer Raman Panwar
Senior DTP Designer Harish Aggarwal
Pre-production Coordinator Tarun Sharma
Pre-production Manager Balwant Singh
Jacket Designer Juhi Sheth
Senior Jacket Designer Suhita Dharamjit
Senior Jackets Coordinator Priyanka Sharma Saddi
Publishing Director Georgina Dee
Managing Director Liz Gough
Art Director Maxine Pedliham
Design Director Phil Ormerod

First published in Great Britain in 2025 by
Dorling Kindersley Limited
20 Vauxhall Bridge Road,
London SW1V 2SA

The authorised representative in the EEA is
Dorling Kindersley Verlag GmbH. Arnulfstr. 124,
80266 Munich, Germany

Copyright © 2025 Dorling Kindersley Limited
A Penguin Random House Company
10 9 8 7 6 5 4 3 2 1
001–345656–Nov/2025

All rights reserved.
No part of this publication may be reproduced, stored in or introduced into a retrieval system, or transmitted, in any form, or by any means (electronic, mechanical, photocopying, recording, or otherwise), without the prior written permission of the copyright owner. DK values and supports copyright. Thank you for respecting intellectual property laws by not reproducing, scanning or distributing any part of this publication by any means without permission. By purchasing an authorised edition, you are supporting writers and artists and enabling DK to continue to publish books that inform and inspire readers. No part of this publication may be used or reproduced in any manner for the purpose of training artificial intelligence technologies or systems. In accordance with Article 4(3) of the DSM Directive 2019/790, DK expressly reserves this work from the text and data mining exception.

A CIP catalogue record for this book
is available from the British Library.
ISBN: 978-0-2417-2364-7

Printed and bound in China

www.dk.com

This book was made with Forest Stewardship Council™ certified paper – one small step in DK's commitment to a sustainable future. Learn more at www.dk.com/uk/information/sustainability